Pesticide Regulation Handbook

A Guide for Users

Jan Greene
Health and Medicine Reporter
Las Vegas Review-Journal
Las Vegas, Nevada

CRC Press
Taylor & Francis Group
Boca Raton London New York

CRC Press is an imprint of the
Taylor & Francis Group, an **informa** business

First published 1994 by CRC Press
Taylor & Francis Group
6000 Broken Sound Parkway NW, Suite 300
Boca Raton, FL 33487-2742

Reissued 2018 by CRC Press

© 1994 by CRC Press, Inc.
CRC Press is an imprint of Taylor & Francis Group, an Informa business

No claim to original U.S. Government works

Library of Congress Cataloging-in-Publication Data

Greene, Jan.
    Pesticide regulation handbook: a guide for users / Jan Greene.
        p.   cm.
    Includes bibliographical references and index.
    ISBN 0-87371-967-0
        1. Pesticides—Safety regulations—United States.   2. Pesticide applicators
    (Persons)—United States.   I. Title.
    KF3959.G74   1994
    344.73'04633—dc20
    [347.3044633]   94-10474

A Library of Congress record exists under LC control number: 94010474

ISBN 13: 978-1-315-89636-6 (hbk)
ISBN 13: 978-1-351-07546-6 (ebk)

Visit the Taylor & Francis Web site at http://www.taylorandfrancis.com and the
CRC Press Web site at http://www.crcpress.com

About the Author

Jan Greene is the health and medical reporter for the Las Vegas Review-Journal, the largest daily newspaper in Nevada. She covered regulation of the pesticide industry between 1986 and 1990 as a reporter for Chemical Regulation Reporter, a weekly newsletter out of Washington, D.C. for which she also wrote two books on pesticide regulation. She has specialized in environmental and health issues during her 12-year career in journalism, and received a bachelor's degree in communications from California State University, Fullerton, in 1983.

TABLE OF CONTENTS

Pesticide Regulation Handbook

A Guide for Users

Chapter 1

INTRODUCTION TO PESTICIDE REGULATION

There was a time, not so long ago, when all a pesticide applicator had to do was read a product's label to know how to apply it safely and what shouldn't be done with it.

Those days are gone. While the pesticide label is still considered the law by federal and state pesticide regulators, much more of the burden for knowing new rules and regulations is being placed on the applicator.

Understanding and interpreting complex and sometimes booklet-length pesticide labels has never been easy.

But now new federal programs to protect endangered species and groundwater from pesticide contamination will require users to go to a pesticide dealer or Cooperative Extension advisor to find out how to comply.

And in many communities, the local city or county government has added another layer of regulation by requiring signs and other notification of applications.

Meanwhile, there is a trend toward requiring applicators to be better trained before they can go out into the field. At the same time, legislators are interested in expanding the universe of people who need to be certified to use restricted-use pesticides, a list that grows longer all the time. And state governments are moving toward requiring certification of people who use all pesticides for a living, not just the restricted ones.

What all this means is that following the rules is no longer a simple matter of reading a label, a task that in itself has never been easy and isn't likely to get easier anytime soon.

More and more these days, it's up to the pesticide user to pay close attention to what's legal and what's not.

Pesticide users themselves are getting more diverse. About 70 percent of pesticides sold in the United States are used for the production of food, including that produced in greenhouses. The rest are used for structural pest control, by landscape gardeners, golf course superintendents, park maintenance workers and public health specialists who kill mosquitos or other disease-bearing bugs.

Traditionally, regulators have kept their focus mainly on farmers because they were using most of the pesticides. But other users are getting more scrutiny these days, such as lawn care companies. That's because there is public concern about the chemicals placed on their or their neighbors lawns.

Also, given the pesticide-of-the-week syndrome, there's no telling when the next chemical or applicator group will be under fire.

So, now more than ever, it's important that applicators of all kinds of pesticides in all kinds of settings be aware of the regulatory system they work under and what motivates it. That makes it easier to be safe, be prepared for the next crisis, and be able to tell customers or other members of the public how pesticides are tested and controlled in this country.

This volume will attempt to paint in broad national terms the kind of regulatory issues coming down the pike. It will also give a brief overview of the federal pesticides law, how it works and how the states carry out its mandates.

There are results from a first-ever 50-state survey of state certification and training regulations. It shows that state programs vary widely in terms of who must be certified and how much they have to know to do so.

Following that is a description of California's pesticide regulatory program, which many believe is the most comprehensive and stringent in the world and the model for what many other states would like to do in the future.

Another chapter talks about why the public is so concerned about pesticide use, and how that drives politicians to write more laws. But it also points out some poisoning incidents that prove there's some substance behind all the worry.

This book also briefly addresses the scientific and technical issues involved in the safety of pesticides, to give you a sense of why these issues often become mired down in talk of parts per million. Often the scientists themselves can't agree about what's risk and what's not.

There's also a chapter on labels: an example of one and some pointers about how to read one and how to recognize symptoms of pesticide poisoning.

We also talked to a few of the people involved in pesticide regulation and offer their thoughts on some of the issues confronting pesticide users these days.

Finally, there's a list of who to contact in your state to get started on the certification process or to get more information about a regulation mentioned here.

It is hoped that this volume will provide some valuable background and context for people facing the increasingly complex world of pesticide regulation.

Chapter 2

THE SYSTEM

Pesticides are regulated under a system in which the federal government makes sure pesticides themselves are safe to use, while state governments make sure applicators are trained and actually use the chemicals properly.

This was set up under a 1972 law called the Federal Insecticide, Fungicide and Rodenticide Act. The first pesticide law was actually put in place in 1910, but all it did was attempt to ensure that consumers got pesticides that actually worked.

The law has been changed a number of times, each time making it a little bit more comprehensive and stringent.

For instance, in 1948 it was expanded to register and label products and require that they include safety warnings. The law wasn't especially stringent, though, because a manufacturer could market its product even if the U.S. Department of Agriculture (which was then in charge of pesticide regulation) turned down the registration.

It wasn't until 1972, a big year for new environmental laws, that Congress started looking at pesticides as a safety risk. That's when FIFRA was substantially rewritten to set up a system in which the new Environmental Protection Agency would take over regulation of pesticides and make sure the benefits of using these chemicals outweighed the health and environmental risks.

The law was also rewritten twice during the 1980s to update the scientific data on the health and safety of pesticides that had already been registered over the years, but we'll get to that later.

Basically, FIFRA regulates pesticides by stating that no such chemical can be used in the United States without being registered with the federal government. It uses the term "pesticide" universally, to include anything that kills bugs, plants, fungus or anything else. The actual definition says a pesticide is a substance or mixture of substances intended for preventing, destroying, repelling or mitigating any pest, or intended for use as a plant regulator, defoliant or dessicant.

The law defines a pest as an undesirable insect, rodent, nematode, fungus, weed or any other form of terrestrial or aquatic plant or animal life or virus, bacteria or other microorganism.

REGISTRATION AND REREGISTRATION

Pesticide manufacturers must go through a lengthy process not only to create their products in the laboratory, but to test them thoroughly and jump through EPA's regulatory hoops to get a product registered.

Most of the 20,000 pesticide products now on the market were first tested and registered 20 or 30 years ago. The industry estimates it would take $10 to

Table 1

Basic Data Requirements for a New Food-Use Pesticide

At the present time, data from the following tests must be submitted to EPA by a manufacturer prior to registration:

Chemistry:

list of ingredients
description of manufacturing process
discussion of formation of impurities
physico-chemical properties
residue studies
metabolic studies
analytical methods
results of analytical procedures

Environmental Fate:

hydrolysis
leaching
terrestrial dissipation
photodegradation
soil metabolism
rotational crop study

Toxicology:

acute oral
acute dermal
acute respiratory
eye irritation
chronic toxicity
subchronic oral toxicity
reproduction and fertility
metabolism
mutagenicity
birth defects
carcinogenicity

Ecological Effects:

aquatic, acute toxicity
avian, dietary & acute oral

$30 million and 5 to 7 years to get a new pesticide on the market now.

The EPA has a whole battery of tests that are required before a product can be "registered" as a pesticide that can be used in the United States. Among the tests are toxicology — whether the product harms the body in an acute, or immediate way; these tests are usually done on laboratory rats and mice, although sometimes they are conducted with dogs or monkeys.

There are also cancer tests, also involving animals, and in some cases environmental tests to see if the product would harm birds, fish or any other part of ecosystems.

The EPA is moving toward requiring additional tests called neurotoxicity tests that determine whether a chemical causes nervous system problems. So far, only certain classes of chemicals known to prompt blurred vision, headaches and other problems have had to undergo those tests; EPA eventually wants all pesticides, new and old, to undergo them. One problem has been that the symptoms of neurotoxicity are so subjective: how do you know if a rat has a headache? Science marches on.

One complaint the industry has had about all these tests is that the standards change constantly. For instance, the first pesticides to undergo these requirements in the early 1970s were tested using the current scientific methods of the day. But now, science has moved along so quickly that those tests are no longer considered scientifically valid. Perhaps the laboratory workers didn't use enough mice, or they didn't take detailed notes or — in the case of one highly controversial chemical, they used vegetable oil to administer the chemical to mice, and the oil has since been shown to alter the test results.

That's what reregistration is all about. All the 20,000 different pesticide products — containing more than 700 different active ingredients — that were tested using 1970s technology are now obsolete and have to be retested at a cost of millions of dollars to their manufacturers.

Most pesticide makers don't complain too much about the fact that reregistration exists — they figure it's worth it to keep the public satisfied that their products are safe. But they are frustrated with the idea that, conceivably, once this set of testing is done (hopefully sometime in the 1990s), will it all have to be done again according to 21st century standards?

EPA officials acknowledge they don't have an answer. At this point, however, they're just trying to get through the reregistration process that Congress required EPA to begin in 1978 and to hurry up with in 1988.

EPA has been trying to get the work done for years, but has been stymied by lack of manpower, cut during the Reagan years, and changing scientific standards.

Critics argue that EPA's real problems were more bureaucratic in nature: computer systems that weren't up to handling that much material; constant turnover in the federal government that meant new people had to be trained to review the tests according to the pesticide program's arcane standards; and just the basic inertia that sets in at any large agency where no one is really accountable for the results.

Whatever the reasons, the hurry-up law passed by Congress in 1988 was supposed to help some of those problems. Congress threw some more money at EPA and gave it some strict deadlines to require companies to do the testing or lose their registrations, and to review the paperwork and make some decisions.

As the bill was being signed, EPA had completed three reregistrations out of nearly 700 active ingredients, and was told by Congress to be done with all of them by 1997.

At the beginning of 1993, EPA had a grand total of 28 done. At that pace, EPA officials acknowledged, there's no way they can get done by a 1997 deadline.

Complicating matters is the fact that a frighteningly large percentage of the tests being conducted to fulfill the new requirements weren't acceptable — they just didn't meet EPA's standards for a well conducted test.

EPA is expected to go before Congress and ask for more money to hire people to go through the huge pile of tests. Officials will probably ask for reregistration fees — paid by manufacturers — to be increased.

Even if the official review of many pesticides isn't done, the reregistration program has had a definite impact on the marketplace.

Products are likely to disappear as manufacturers drop pesticides that just aren't worth the cost of additional tests. Of the 613 ingredients identified by EPA as needing to be reregistered, 379 of them have had a manufacturer or group of makers agree to do the testing. Another 206 have not.

Big-ticket items like Roundup aren't about to go by the wayside. But growers of specialty items, called minor crops, began to lose products designed specifically for them.

That has caused considerable concern among those growers and prompted Congress to consider ways to get the studies done by someone other than the manufacturer or to get EPA to relax its standards for certain low-risk minor use pesticides.

According to a June 1992 General Accounting Office report on the problem, more than 40 percent of the nation's $70 billion agricultural sales were from minor crops — vegetables, fruits, nuts and ornamentals.

The GAO found that a government program set up to conduct tests to save important minor use pesticides was seriously underfunded and that the U.S. Department of Agriculture wasn't doing much about it. Ultimately, Congress will be asked to make this pesticide research money a priority at the same time it is trying to reduce the federal budget deficit.

OTHER FIFRA PROVISIONS

Besides registering pesticides, the other major feature of the FIFRA is the emphasis put on the label. EPA likes to say "the label is the law." That means it is illegal to do anything that is prohibited on a pesticide label. That makes some pesticide labels amazingly long — they actually turn into booklets. They explain the potential health effects of the product, how to use it safely, whether to wear protective gear when using it, when it's safe to go back into a field, etc.

One of the key things to look for on a pesticide label is whether the product is registered for "restricted use." If it is one of the approximately 100 restricted use pesticides, that means special care must be taken in using it; also certain additional requirements kick in. For instance, most states require extra training and certification for anyone using restricted pesticides. And the federal law requires that if someone is using a restricted-use pesticide as a business, they must be certified.

Once a pesticide has been registered and a concern comes up about whether it is safe, EPA has options to study it and decide whether to take it off the market. It can begin a special review, an intensive examination of a pesticide's risks and benefits.

If EPA decides it must take some kind of action against a pesticide — either with or without a special review — it can cancel a particular use or its entire registration. However, that can be challenged through a hearing if the manufacturer requests one. And the cancellation must be sent to USDA and an EPA committee called the Scientific Advisory Panel before it is done.

During the cancellation process the pesticide may still be sold and used. That is, unless EPA finds there is an imminent hazard and suspends its use while the pesticide's status is under review. Even if EPA finds there is an immediate hazard requiring an emergency suspension, it must balance the benefits of use, something that has made it difficult for the agency to use that provision. EPA officials have asked that it be changed so the government can more easily act against problem chemicals.

Once a pesticide is pulled from the market, EPA is required under FIFRA to collect all of that product from the marketplace and get rid of it. That has turned into a difficult, time-consuming and expensive process for EPA. Only a handful of pesticides have ever been canceled.

For instance, EPA canceled registrations of dinoseb in 1986. It guessed the government would end up spending tens of millions of dollars to collect, store and incinerate about 4 million gallons of dinoseb. On top of that, EPA used to have to pay manufacturers for any unused stocks of a canceled pesticide, which was expected to add up to about $40 million for dinoseb alone. That provision was changed by Congress in 1988, when it also changed the storage and disposal provisions for canceled pesticides so that the manufacturer shares the burden and cost.

TOLERANCES AND FOOD SAFETY

There is another law that regulates some pesticides: the Federal Food, Drug and Cosmetic Act. This law, carried out both by EPA and the Food and Drug Administration, establishes "tolerances" — levels of pesticide residues that may remain in food.

This is actually the area of hottest debate in the pesticide world these days because of a section of the FDCA known as the Delaney Clause. It is now being debated whether the 30-year-old sentences in that clause of the law were originally intended to mean that there could be absolutely no risk from traces of a carcinogenic pesticide in food or, as the federal government and industry have argued, there can be a tiny amount that poses just a "negligible," or insignificant risk of cancer.

Pesticides can be registered under FIFRA for use on a food or feed crop only if a tolerance (or exemption) is first granted. EPA has approved about 300 pesticides for food uses, according to a 1991 EPA document, and about 200 of them are in common use in the United States.

The FDA enforces the tolerances set by EPA. It has a program for monitoring a small percentage of pesticide residues in food and determining how much risk is getting through to the public. The agency has reported that very little leftover pesticide gets through to the food, though FDA has its critics (see next chapter).

CERTIFICATION AND TRAINING

Actually, the federal pesticides law doesn't say a whole lot about how pesticide applicators are regulated. The bulk of the law talks about how pesticides are tested and registered, and only a small section addresses how to be sure the people who use pesticides are properly trained.

Basically, that section simply requires that anyone using a restricted-use pesticide — one on a list of about 100 chemicals considered particularly hazardous — be certified. It's up to the individual states to figure out how to certify applicators and what kind of training, if any, is required for certification.

That limited discussion of applicators in FIFRA may change if some people in Congress have their way. Some of the proposals under consideration in recent years would require applicators of all types of pesticides, not just the restricted ones, to be certified. Also, some would like to increase the training of certified applicators.

That includes a couple of major industry organizations that represent many of the pesticide applicators in the country. The National Pest Control Association, whose members do mostly structural pest control, has taken a stand in favor of drawing all professional pesticide applicators of non-restricted, or general use, pesticides into the regulatory fold.

The trade group also supports mandatory verifiable training for technicians who work for companies that apply pesticides. And in-house pesticide applicators, such as janitors, custodians, groundskeepers and building managers, should also be included in training and certification requirements, the group argues.

While those positions might seem to increase the regulatory burden on the industry, the pest control association believes that strengthening the requirements for pesticide applicators will improve the industry's image in the public's eyes. Also, the incidents of misapplication that result in harm to customers of pest control firms often happen because the workers were not adequately trained. Those incidents make the entire industry look bad.

The Golf Course Superintendents Association has also taken a strong stance on the training issue, creating an extensive continuing education curriculum for its members. The association has its own certification program. Before anyone can even enter the program, they must have five years' experience as a golf course superintendent and have completed a year of college course work or have 10.5 continuing education units from the association's seminars. The group even offers specialized training. For instance, an integrated pest management specialization includes seven seminars on topics such as "environmental considerations in golf course management" and "scouting, sampling and monitoring golf course pests."

The golf course superintendents group also supports state certification, and notes that 97 percent of its members surveyed had at least one state certified pesticide applicator on staff.

That state certification program started in 1974 when EPA began carrying out its mandate under 1972 amendments to FIFRA. They required that EPA set up standards and develop training materials for the education of pesticide users in the certification programs and that states would carry them out.

Within two years, all but two states were carrying out those programs. The two that don't, Nebraska and Colorado, have their certification and training programs carried out by the EPA regional offices in their areas.

Among the basics of the certification program:

—There are two types of basic pesticide users: private and commercial. Private applicators are essentially those who use pesticides on their own farm property, commercial applicators are those who apply as a business.

—Certification is required only for restricted-use pesticides. General use (everything else) does not require it.

—Anyone being certified has to be trained and prove their competency. Commercial users must do that with written tests, private applicators do not.

—All states must adhere to the basics of the EPA rules but can be more stringent than the federal government if they so choose.

A decade after the certification requirement was created, more than 2 million private and commercial applicators had been trained across the country, resulting in about 1.2 million people being certified by states.

But also by that time it was becoming clear that reform was needed to fill in some gaps. In 1985 EPA put together a task force to take a closer look at some of those issues.

The biggest problem was how to interpret FIFRA's requirement that restricted-use pesticides be used only by certified applicators or by someone under the "direct supervision" of a certified applicator. Not only did some see that as a major loophole in the restricted use rule, but different states were applying it in widely varying ways. Some applied a "line of sight" rule requiring that the certified person be able to watch the work directly. Others said the certified applicator should be within reach by telephone and others didn't address it at all.

The task force recommended that EPA use a section already in its regulations to clarify the issue with specific pesticides. That part of the rules says that restricted-use pesticides can be used: 1) only by a certified applicator; 2) by a competent, non-certified applicator working in the actual presence of a certified applicator; or 3) under the direct supervision of a certified applicator who has provided verifiable instruction to, and is in control of, the competent, non-certified person. Because the judgment for which criteria to use was left up to the applicator, based on his or her judging the hazard of the situation, there has been great variability in how the requirement is actually carried out. The task force suggested that EPA give more guidance on product labels which of the three criteria apply to use on that pesticide.

The task force also recommended that training materials include information on liability and insurance; professional responsibility of the applicator to use pesticides properly; and dealing with public concerns. Training materials should also be updated to include new information on scientific review of chemicals, Groundwater contamination, endangered species concerns, disposal and proper calibration of spray equipment, the group said.

The task force said that private applicators — usually farmers — should be tested, something that is not required under FIFRA.

Another weakness identified by the group was recertification — it noted that nowhere in the law does it say that word to refer to certified applicators renewing their ability to use restricted-use pesticides, but the law's intent is clear that anyone wanting to maintain that right should undergo continuing education. The result has been a hodgepodge of state rules on how and when applicators are recertified. EPA should toughen that part of the rules, the panel said.

Also confusing among states is the fact that some also require licensing separate from certification. Licensing is usually just having a pest control business pay a fee to do that business in the state. States should clarify the difference between licenses and certification, the task force recommended.

Although the task force completed its work in 1985, this book's 50-state survey (including the District of Columbia) shows that many of its concerns are still problems.

For instance, many states' rules were unclear about exactly what licensing meant and what certification meant, and whether one required the other.

About half the states don't require private applicators to take a test to be certified. In some cases, all they have to do is show an extension agent they've read a book.

In some states, extensive training materials are provided on a wide array of topics, and applicators must take courses to keep abreast of new information. But in a large number of states, the training program hardly goes beyond the bare minimum required under FIFRA back in 1972.

EPA officials wrote up some changes to their rules in response to the 1985 task force report, but because certification and training has not been a top priority within the agency, it still had not been approved by early 1994.

In that proposal is the major change recommended by the task force: that EPA be clearer about the circumstances under which a non-certified applicator can use a restricted-use pesticide.

The proposal would establish three types of use levels: 1) use only by a certified applicator; 2) direct supervision by a certified applicator who is required to be on site at all times and available at the point of use within five minutes; and 3) direct supervision by a certified applicator who is not required to be on site. EPA would assign one of these levels to each restricted-use pesticide based on its potential hazardousness.

The proposal would also establish new categories for private applicators: fumigation of soil and agricultural products, chemigation and aerial application.

It would add new categories and subcategories for commercial applicators.

Some states have already established their own categories for private and commercial applicators.

Another provision of the proposal would make the two state programs run by EPA slightly more stringent. In Colorado, EPA's regional office certifies private applicators; in Nebraska it runs both the private and commercial programs.

Interestingly, because EPA cannot be more stringent than itself, its regional offices are restrained from requiring many of the things states require in their certification programs. So in Colorado, private applicators are sent a home study course with a multiple choice, true-false test. They send the test back and if it passes, the applicant is certified. If the applicant didn't get 80 percent right, the test is sent back for a second time. But EPA doesn't have authority to turn down certification the second time even if the applicant doesn't pass the exam.

Resources are also a problem. EPA's regional office certifies 13,000 private applicators every four years through the process, and it can't charge a fee for it. One person is assigned to carry out the entire program.

Q & A Jeff Pifferini, superintendent of the Spanish Bay golf course, a five-year old course and one of four at the famous Pebble Beach resort in Monterey, California.

Q. Are the regulations difficult to understand and are they getting more complex to follow?

A. Sure, it's getting more complex every day. (California) and Massachusetts are the two toughest states, and the rest are going to end up the same way.

But I've gone through all the classes. Of course, knowing everything is one thing, doing it is another.

I've got a certification. It's getting to the point where everybody's going to have to have one.

Q. Is that difficult?

A. Well, you've definitely got to study and you've got to pass a test. A lot of people take it 2 or 3 times. You have to know all the laws and regulations.

And then you have to take another test in whatever field you're in. So it takes two individual tests. They don't just give them away.

Q. How many people work on your course? Do all of them need to be certified?

A. I've got 24 total (on the maintenance staff). I've got three guys who are gonna take the test coming up. By law you only need one (certified applicator) on the course.

Q. Is there more pressure running a course in as famous and environmentally sensitive a place as Pebble Beach?

A. You've got a lot more eyes looking at you. You just play the game, follow all the rules and regulations they want and everything's fine. If everybody just followed all the rules it'd be fine. It's when you don't use common sense that you get in trouble.

I took the national environmental and pesticide courses (offered by the Golf Course Superintendents Association). They teach all the groundwater stuff, about runoff into ponds and groundwater. Spanish Bay's situation is we have no ponds or lakes and we have a wetland riparian area at one end of the course that drains into and another end and nothing in the middle. They did it for that purpose.

Also, the amount of chemicals we use and the dilution rate is so minute that what you have under your sink or I have under my sink is twice as strong. It's really kind of amazing what you put on cleaning tile is so much stronger. If you follow the label or go to the light end of it ... you don't want to damage the turf.

Q. Do you have any PGA tournaments on your course?

A. No, that would be too much of an environmental impact. We're just a resort golf course.

Q. Do you get many questions from golfers or the public about the pesticides used on the course?

A. There is a binder that has the (material safety data sheets) in it that they can see. It's available to them if they ask. But not really, I don't really see it that much. We don't do as much spraying as some other courses. We have a pretty fescy turf, it's pretty tolerant of disease or drought, which is pretty good.

Q. How do you keep up with constantly changing regulations?

A. I just keep taking courses. And then the ag commissioner stops by and fills us in too. If there's anything new he brings it by. It's a pretty-good watched operation.

There haven't been any accidents of any kind in the past few years. Of course, if you have a broken line that shoots at somebody then everybody has to know. The doctor has to know, and he has to let the ag commissioner know. It's pretty wide open. It's not a hidden thing.

Q. What kind of protective equipment do your applicators wear?

A. They always wear protective gear, no matter what they spray. You need it all for any handheld operation, but we require it even when they're riding the tractor. You're better safe than sorry. So it costs a little bit more to have all the equipment. It's works out a lot better that way.

Chapter 3

PESTICIDES, POISONING AND THE PUBLIC

Pesticides: the word conjures up different things to different people. To professional pesticide users it means a useful tool to do a job: save a farmer's crop from fungus, a homeowner's house from termites, a community from disease-carrying vectors.

But to most of the public, the P word means poison, cancer, big business making a profit at the expense of the public's health.

As in most things, the truth lies somewhere in the middle. Over the past 10 years, that truth has become more muddled as pesticide regulation has become more and more politicized.

Although attention has always been periodically focused on pesticides every time there was a health scare — known within EPA as "the pesticide of the week syndrome" — the peak of national concern about the issue seemed to come around 1989, when the apple growth regulator Alar was the topic of a "60 Minutes" broadcast — twice — and actress Meryl Streep performed before an unusually packed Senate hearing to express her concern about feeding apples to her children.

The Alar scare seemed to spotlight the problem with regulating a chemical in the food supply under the American system of government: the incredible complexity of not only the science behind the risk from Alar's breakdown product, UDMH, but how residues are measured in foods by EPA and the complicated formula it uses to assess risk. Throw in politics and the press, and the "truth" of health risks from eating apples was hopelessly obscured for the average person.

It would be easy to dismiss the Alar incident as a case of pesticide critics manipulating the media because, in part, it was. But that would ignore the core issues involved in that and every other health concern identified by EPA about a pesticide — remember, these are chemicals created to kill things, and the scientific community is only just beginning to understand the long-term health effects of many chemicals.

That there really is something to fear from pesticides is tragically reemphasized with the poisoning of a farm worker. Or with an incident such as the 1987 poisoning deaths of an elderly couple in rural Virginia who came back home after their home was sprayed for termites. They both died, one a few days after the other.

Obviously, these kinds of incidents should always be kept in mind by anyone using pesticides professionally. It's easy to get caught up in the defensive backlash of the industry against what it sees as unfair criticism.

More difficult are the not-so-obvious cases, such as when a pesticide user is told not to use a product anymore because it caused tumors in laboratory animals. At this point, that's the only way the scientific community has to warn us of danger.

Scientists offer that information, by saying, "Okay, here's how many tumors we found, you decide whether you want to be exposed to this.'' It's one thing to be able to make that decision for oneself, as a pesticide user can do, theoretically. But for the people eating the food, living in the houses, playing the golf courses and picnicking on the lawns where these products are used, often they don't have a choice, because they don't have the information. And that makes people, especially Americans, very uncomfortable. Thus the trend toward "right to know" and posting regulations. More and more, people want more control over the chemicals they are exposed to.

So for the user, knowing what the label says and what it means is more and more vital, not just for your own safety but for the comfort of your customers.

Of course, there's often a third party involved in this decision-making process: the politician. By the time a politician — on any level of government — gets involved in a pesticide issue, you can bet there's been a crisis that's prompted calls from constituents.

This is especially true in Congress, where pesticide issues are understood by a handful of members, who often have such a full plate that there has to a be a full-on emergency to get anything done. In the mid-1980s, even though EPA was years late and obviously incapable of reviewing all the older pesticides on the market, as Congress ordered it to do, Congress couldn't manage to get a bill passed to help the agency do the job until the Alar scare woke everybody up. In Washington, it isn't enough that something needs to be done, there has to be the political will to get it done.

Which is one reason why regulation of pesticides might seem fragmented to the user. You may well ask, Why isn't anyone doing anything to save pesticides used on minor crops? Why doesn't EPA simplify labels? Why don't federal agencies such as USDA and EPA get together and coordinate their regulations?

Some of these issues, such as better labels and minor uses, have been floating around Congress and EPA for a decade or more. But there hasn't been enough momentum or money to resolve them.

One problem with the piecemeal way things get done at the federal level is that rarely does anyone stop and take an overall look at the whole pesticide regulatory program. Some scientists are urging EPA to do just that — step back and question why so much money is going toward the assessment of minuscule health risks from pesticides in food when thousands die from smoking each year. For that matter, plenty more farmworkers and other pesticide users are at direct risk of poisonings and long-term illness than anyone munching an apple. In an era of dwindling resources, EPA may have to start taking a closer look at what it spends money on.

Of course, if the American people decide they want to spend their tax dollars making sure there's not an invisible, long-term health risk in the food they eat, who's to say they shouldn't? That's the American way.

Chapter 4

ISSUES IN REGULATION

As the law stands now, you don't have to do anything special to use a general-use pesticide — no training, no tests. That's only for people using restricted-use products, those 100 or so active ingredients considered particularly hazardous.

But things are fast changing, so that being certified is an important step for anyone handling pesticides, even if they don't think they'll use the more hazardous products. One reason is that rules are fast becoming more stringent so just about all pesticide users will likely have to get training and take tests.

But another good reason to get certified is to keep up on new rules that are important but hard to find out about without going through official training courses.

For instance, new maps are coming out in many parts of the country that show to the exact acre where certain pesticides can't be used because they may harm an endangered plant or animal.

There are also new ground water protection programs coming from state governments with rules that may not appear on the label.

And EPA is working on a big new program to rework how pesticide containers should be stored and thrown away. Already the program — required under 1988 changes to FIFRA — is late, and there's no saying when it might become reality.

The agency has to issue regulations on procedures for removal and rinsing of pesticides from containers; safe use by eliminating splash and leakage; and safe disposal, refill and reuse. EPA is also supposed to look into ways to encourage recycling of pesticide containers.

So it's clear there are changes on the horizon that will directly affect most pesticide users in the country.

Many user groups, such as those representing pest control companies, golf course superintendents and professional landscapers support certification and training for all pesticide users, partly to improve public confidence in their industries and partly to ensure the people who are doing the work are competent to do it.

Becoming competent these days, though, is no longer a simple matter of reading a label, putting on a mask and calibrating a sprayer. There are many emerging issues that will require constant vigilance by applicators, both neophytes and those with decades of experience.

This chapter will explore some of those emerging issues.

LOCAL POSTING LAWS

In the past several years people have become much more worried about pesticide use in their communities, whether it's the next door neighbor hiring a lawn care company or a public building that's been sprayed.

That concern, and the general public perception that no one in the federal government is doing anything about their worries, has led to more and more local action by cities and other jurisdictions.

The trend was cemented even further in mid-1991 when the U.S. Supreme Court said the federal pesticides law does not pre-empt local jurisdictions from writing such rules.

The court opinion has prompted a lot of action among pesticide user groups, who want Congress to pass a law clarifying that pesticide restrictions should come primarily from the federal government.

In fact, a congressional subcommittee in 1992 approved legislation to do that, though it didn't get any farther than that. The bill would have forbidden states to allow local units of government to regulate pesticides, including posting and notification rules, and would have prevented states from moving faster than EPA with requests to manufacturers for pesticide safety data. In other words, states like California could no longer require pesticide makers to submit scientific studies on the health effects of pesticides if EPA hadn't already asked for that test.

The first part of that legislation is what pesticide user groups were pushing for that year, but they now say the effort in Congress is less urgent because many states are passing laws that keep lower jurisdictions from restricting pesticide use or requiring the posting of signs during applications.

Those favoring local control over posting regulations argue that they are needed because the public often doesn't get enough information about toxic chemicals in the environment. The National Coalition Against the Misuse of Pesticides, for instance, has testified that local rules on pesticides are no more burdensome than individual city building codes or protection of local water supplies.

The environmental advocacy group contends that people coming in contact with pesticides that have been applied to a neighbor's lawn, a golf course, a park or any other public place have no way to know what kind of chemical is there or if they should be worried about it. Some people may be sensitive to those chemicals or become sensitive after repeated exposure, NCAMP says.

According to NCAMP, there may be real risk to low-level exposures to chemicals used in the lawn care industry. The group says that of the 34 most commonly used chemicals in lawn care, 10 are cancer causing, 12 cause birth defects, seven have reproductive effects, 20 are neurotoxic (they damage the nervous system), 13 cause liver or kidney damage and 29 are sensitizers or irritants.

These chemicals can also cause environmental damage, with large bird kills on golf courses or other large expanses of grass documented more than 36 times in recent years.

Lawn care pesticides can also end up in groundwater. The single pesticide found most often in the EPA's national survey of pesticides in groundwater was

Dacthal, an herbicide used commonly in lawn care, not crop production.

Supporters of local control also contend that EPA is not doing a good enough job testing or monitoring pesticides, so city and county governments are an important backstop to ensure the public is protected.

On the other end of the spectrum, industry groups contend that pesticides are, for the most part, well tested and carefully used. They fear that local jurisdictions will make unreasonable rules that have no basis in science or fact, and they will cost the industry time and money.

For instance, posting laws typically require that signs of a certain size with certain language on them be placed so many feet apart from one another for so many days ahead of and after a pesticide application. They may also require that neighbors and chemically sensitive people who have previously signed up on a registry be notified several days in advance.

The problem with these laws can be that they delay the process so much an applicator can't respond to an emergency situation. Or that it is unnecessary to spend the time and money to put up signs for neighbors who may not even care about the information.

Instead, groups like the National Pest Control Association say, the information about a pesticide application should be made available to people who request it. And any such rules should be written by state or federal governments to maintain consistency. They point out that there are more than 83,000 units of local government in the nation, so there could theoretically be that many different rules to follow.

The NPCA has come up with the following ways it believes notification of an application should be made:

—Advance information, including a copy of the pesticide label, should be made available to a residential customer upon request.

—In an apartment house, the owner or manager should be given prior notice of the application but it would be up to that person to notify everyone else in the building.

—In public buildings a permanent sign should be put up telling people that the facility is periodically treated.

—For lawn and ornamental applications, a sign should be placed at conspicuous points of entry at the time of application and for 24 hours thereafter, or until the product is dry.

—People who are especially sensitive to pesticides should be notified in advance of an application on adjacent properties.

Another industry group, the Golf Course Superintendents Association, also backs a state- or federal-level standard for notification and posting rules.

It also recommends to its members that they voluntarily post an appropriate warning in a conspicuous place on the course that lists the name and phone number of the superintendent and a statement that chemicals have been used in the past 48 hours. Anyone wanting more information could get it from a material safety data sheet from the superintendent.

The debate over whether the federal government should pre-empt local action on pesticides has gone beyond posting and notification, however. There have even been towns that have adopted their own permit requirements for applicators. In fact, the Supreme Court case that brought all this to a head in 1991 (Wisconsin Public Intervenor v. Mortier) was over the town of Casey, Wisc., having an ordinance requiring a permit for applying any pesticide to public lands, private lands subject to public use or aerial application to private lands. The Supreme Court upheld Casey's right to have its own pesticide permit system.

SEEKING CONSENSUS

This debate raged on in Congress during 1991 and 1992, until a couple dozen states took action to stop their own cities from passing such laws. That has quieted the issue somewhat on the federal level, but many would still like to see a national standard for posting and notification.

That's why the EPA convened a Lawn Care Pesticides Advisory Committee. The panel drew members from health and environmental groups, industry, state government, congressional staffs and various federal agencies.

The task force issued a report in February 1993, and, to no one's surprise could come up with no single answers. In fact, the group split up into factions that issued their own position papers.

The group did come up with summaries of each of the major issues facing lawn care:

Pre-emption

As of April 1992 at least 53 localities in 20 states had enacted some type of lawn care legislation and another 50 localities had proposed lawn care legislation. The panel identified four different major points of view about whether this should be pre-empted at the federal level, but all of those points of view favored the establishment of a national standard. However, some felt it should be mandatory and others urged that it be up to states whether they adopt the standard. Also, some felt that local jurisdictions should be able to exempt themselves from that standard for good reason.

Posting and notification

Many panel members agreed it may be a good idea to some extent; some felt there wasn't enough scientific evidence to show that passersby might actually be harmed by walking by a recently treated lawn.

One interesting question that came up was whether homeowners themselves should have to tell their neighbors when they apply pesticides. It was noted that homeowners apply about two-thirds of the lawn care pesticides used on home lawns and about 43 percent of all turf pesticides.

According to the panel's report, 24 states require professional applicators to provide notification when applying lawn care pesticides to residential sites. They are Maine, New Hampshire, Vermont, Connecticut, Massachusetts, Rhode Island, New York, New Jersey, Pennsylvania, Maryland, Delaware, Ohio, Florida,

Indiana, Kentucky, Illinois, Wisconsin, Minnesota, Iowa, Kansas, New Mexico, Arizona, Washington and Colorado.

Posting

Sixteen of the 23 states require posting of a sign when the lawn is treated. Of these 16, only Connecticut requires homeowner posting.

The panel told EPA it should think about whether it wants to develop enforceable posting and notification standards or just guidelines, and what kind of information should be required. The agency should also consider whether homeowners should be subject to these kinds of rules as well.

Registries

According to the task force report, eight states maintain registries of people who are to be notified when professional applicators apply lawn care pesticides within a certain distance from their homes. They are Connecticut, Colorado, Florida, Maryland, Louisiana, Pennsylvania, West Virginia and Washington.

Some are so-called open registries, while others require the person registering to provide medical certification of their chemical sensitivity. The task force members actually favored the open kind, noting that it may be difficult to get the medical certification, it may have an effect on someone's ability to get health insurance, and it may infringe on their medical privacy.

There are also questions about whether people should have to pay a fee to get on the registry.

Potential action

This issue is not likely to go away soon, given the wide gaps that divide those debating it. But the trend certainly seems to be toward more information being available to people about pesticides used on lawns, whether that is required by a local government or mandated by Congress.

Meanwhile, EPA is looking at ways to resolve the questions about whether people are really being hurt by these types of pesticides. The agency is trying to come up with better ways to quantify the benefits of lawn care pesticides and at the same time how serious the health risk is to residents and birds from pesticides used on turf and golf courses. EPA is also looking into how it can monitor residential and children's exposure to lawn care chemicals.

WORKER SAFETY

The EPA in mid-1992 issued a long-awaited new rule to improve the safety of workers handling agricultural chemicals. This rule is sometimes called the farmworker safety rule or the 1992 worker protection standard. Importantly, the standard is limited to people who work in:
—agricultural fields
—greenhouses
—nurseries and
—forests.

It does NOT apply to people:
— in government-sponsored public pest control
— working with livestock
— who garden, do lawn care or apply chemcials around homes
— doing structural pest control
— handling vertebrate pests
— putting out attractants or repellants in traps
— doing post-harvest applications or
— researching unregistered pesticides.

Because the bulk of pesticides used in the United States are used by agricultural workers, this rule was expected to affect 3.9 million people.

Among the major provisions are that employers provide pesticide handlers and workers with cleaning supplies for washing or decontamination and that emergency transportation be provided if someone is hurt by a chemical. It also adds new time periods for when someone can re-enter a treated area and requires personal protection equipment to be used.

Many product labels already include requirements for re-entry intervals and protective equipment. But the new worker protection standard would expand those regulations.

Everyone must comply with the rule by April 15, 1994.

Specifically, here's what it does:

Centrally Located Information

A poster must be displayed containing information on worker protection, including the location of the nearest emergency medical facility. There must also be information about each pesticide application made on the establishment, including the location and description of the treated area, the product name with EPA registration number and active ingredients, time and date of the application and restricted entry interval for that pesticide.

The information about applications must stay up for at least 30 days after the re-entry interval is over. Workers and pesticide handlers must be told where the poster is located and they must be able to get to it.

Emergency Assistance

In case of a suspected pesticide poisoning, the employer must make available prompt transportation to an appropriate medical facility. The employer must also provide the worker or handler or the medical personnel with information from the pesticide label and how the suspected exposure occurred.

Decontamination

There must be a decontamination site nearby any place where workers are working where a re-entry interval is in effect or has expired in the past 30 days.

Pesticide handlers must always have a decontamination site nearby where they are handling pesticides. (Handlers are considered people who mix, load, transfer or apply pesticides; dispose of pesticides or unrinsed conainers; handle opened containers; flag; clean, adjust, handle or repair contaminated equipment; assist with application; enter enclosed area after use of airborne pesticde before PEL or ventilation criteria are met; enter an area treated with soil fumigant to adjust or remove tarps; perform tasks as a crop adviser during application or a re-entry interval.)

Supplies for washing pesticides from the skin and eyes must be provided within a quarter mile of all workers and handlers, including:

—enough water for washing (water must be of good enough quality and temperature so it doesn't hurt someone using it or swallowing it);

—enough soap and single-use towels;

—and, at handler sites, clean coveralls.

Anyone having to use protective eyewear must have eyeflush water immediately available.

Information Exchange

A commercial handler applying a pesticide on a farm must inform the agricultural employer about what's being applied and must provide information that must be posted, plus whether both oral warnings and posting are required and any other protection requirements on the label.

A commercial handler employer must tell the employees about any entry restrictions.

Entry Restrictions

The rule sets up general restricted entry intervals, meaning the time workers must wait before they go back into a treated area. Many pesticides already have those kinds of restrictions on their labels, and those will be retained as long as they meet the guidelines in the new rules.

For other pesticides, the rule establishes a 48-hour interval for any product that is highly toxic because of dermal toxicity or skin or eye irritation. It is extended to 72 hours in arid areas if the product is an organophosphate and is applied outdoors.

A 24-hour interval is established for any product that is moderately toxic because of dermal toxicity or skin or eye irritation.

A 12-hour interval is established for all other products.

Workers not involved in the application must stay out of the treated area altogether.

There is a whole set of rules for protecting workers who go into the treated area before the restricted entry interval is over. For one thing, they can't go in until four hours after the application and until any label-specified inhalation exposure level has been met. Also, personal protective equipment must be provided and worn. The worker must have a clean place to put and and take off the equipment and have soap, towels and water available when he takes it off.

Training

Unless the worker is already certified as an applicator or a trained handler, each early-entry worker must be trained before working in the treated area. The training must include written or audiovisual materials and be presented in a way the worker can understand.

Notice of Applications

On farms, nurseries and forests, each worker who might enter a treated area or walk within a quarter mile of a treated area during application or a re-entry interval must be warned orally or by posting warning signs at the treated area. In greenhouses, there must be warning signs.

A similar set of rules applies to handlers.

Certain exceptions are allowed under the rules, such as exempting pilots and handlers in enclosed cabs from some of the personal protective equipment requirements.

Also, a grower can request an exception to the re-entry interval by applying to EPA.

Some employers are concerned that the new rules are too general and don't make sense in specific situations.

However, farmworker advocates have been calling for these kinds of overall protections for years, and find the final result wanting in some ways. For instance, the new rules require that warning signs include the word "danger" in Spanish. But still, pesticide labels don't have to be in any language but English despite the fact that a large number of farmworkers are Spanish speakers.

State officials will also have to work out arrangements to handle the training requirements in the new rules. Workers must be trained before their 6th day of entry into treated areas on an agricultural establishment. Some worry that migrant workers will end up getting the same training over and over, each time they move to a new state and a new crop. State officials are considering some kind of card workers can carry to verify they've been trained.

ENDANGERED SPECIES

Another initiative EPA has been working on for several years is protection of endangered species from pesticides. Under the Endangered Species Act, any federal agency must take action to protect those species; once there was evidence that certain rare plants or animals were being harmed by pesticides EPA was forced under the law to do something about it.

In the late 1980s EPA came up with a plan to put a general statement on a pesticide believed to pose a hazard to a particular species and direct users to get more information locally. Then it issued a series of local maps of areas where endangered plants and animals lived and where specific pesticides shouldn't be used. Problem was, many local officials complained these maps were wildly inaccurate, so EPA had to try again.

So it went back and redid the maps, put out fact sheets and allowed states to come up with their own programs if they wished. At last count, 15 states decided they wanted to do so.

What this means for users is that sometime in 1994 a mandatory program will go into effect that requires you to check the pesticide label for any warning about endangered species. Then you must check with Cooperative Extension or the pesticide dealer for specific fact sheets and maps describing the specific restrictions in your area. In some states this process may differ slightly.

EPA is slowly going through its list of several hundred active ingredients to find those that might cause harm to rare species. EPA is doing this through biological consultations with the U.S. Fish and Wildlife Service, which is part of the Department of Interior.

Some states are going ahead with their own plans, in part to preserve the use of certain pesticides in areas where EPA would completely ban them. For instance, North Dakota has adopted its own less restrictive plan that restricts the use of 37 pesticides in 23 counties. But none of the pesticides is prohibited completely in any county, as EPA had proposed to do. State officials said it didn't make sense to ban a pesticide in an entire county if it was only posing a problem in a small area.

The North Dakota plan sets up three classifications for areas based on whether endangered species are definitely present, have been present in the past six years, or have never been recorded as living there. The state sends farmers maps of the restricted areas and allows them an opportunity to appeal a use restriction if they really need to use a listed pesticide in that area.

GROUNDWATER

A problem that has surfaced over the past several years is pesticide contamination of groundwater. EPA has done a number of well water surveys to try to assess the seriousness of the problem, finding in a national survey issued in 1990 that there were 12 pesticides above minimum reporting limits and six were found above health advisory levels.

While many of these were rural wells, and many people expected farm chemicals such as atrazine to be found most often, the actual results were a little surprising: the most frequently appearing pesticide was Dacthal, a herbicide commonly used in lawn care.

The survey also found that nitrates are a more serious threat than pesticides at this time. It estimated that 10.4 percent of the community water systems and 4.2 percent of rural domestic wells contain pesticides above the minimum reporting levels used in the survey.

While EPA has attempted to take direct action based on this information, it has been stymied by the simple geologic fact that groundwater basins are inherently local, making it difficult to regulate on a nationwide basis.

Up to now, pesticides have been dealt with by EPA on a case-by-case basis for groundwater concerns. For instance, when atrazine began to appear in small amounts in many parts of the country where it was used, its manufacturer did a nationwide survey of wells and surface water and ended up voluntarily changing the product's label to reduce the hazard.

But because so many different products are showing up in well water EPA is trying some more comprehensive approaches.

One is a proposed rule that would restrict the use of certain pesticides considered to have the potential to leach into groundwater. The rule would classify as "restricted use" those problem products so that only certified applicators could use them.

The agricultural chemicals industry worries that these nationwide bans on using certain products because of groundwater concerns may be unnecessary if they only pose a problem in certain areas.

But others believe EPA needs to be more stringent about studying and limiting potential groundwater contaminants. For instance, the General Accounting Office said in 1992 testimony to Congress that "EPA has made limited progress in reviewing studies on the potential for these pesticides to leach into groundwater.

"In addition," the GAO continued, "EPA had not fully utilized the regulatory measures it had available to reduce groundwater contamination by pesticides. Such regulatory measures include cancellation of pesticide uses as well as less severe measures such as placing advisories on labels or prohibiting use in specific geographic areas vulnerable to contamination."

In any event, EPA is trying to deal with the issue in a global way with its "pesticides in groundwater strategy." That plan sets up a basic framework for regulation but requires states to come up with the specifics needed to carry it out. Based on the theory that groundwater contamination is basically a local problem, the states would come up with their own plans and have EPA approve them.

Some environmentalists worry this could abdicate the entire issue to states, some of which may not be interested in pursuing strict regulation, or may not be able to afford to.

Still, that's the approach EPA is taking. It is currently reviewing state plans.

The moral of the story? Keep an eye on the label and an ear to the ground for how your state will carry out this new mandate.

RECORDKEEPING

A new rule issued by the Department of Agriculture explains the kind of information users of restricted-use pesticides must keep and provide to farmworkers and medical personnel in the event of a poisoning.

The rule was the result of the 1990 Farm Bill. It wasn't issued until an environmental group sued USDA to force its release. Less than a month after the lawsuit was filed, a proposed rule was issued and in April 1993 a final version put out.

The rule covers all restricted-use pesticides used by certified applicators. The records are needed to form a data base for agronomic and environmental surveys by state and federal agencies and for annual reporting to Congress on the use of restricted-use pesticides. The regulations include a provision for protecting the identity of individual producers.

Some states already have recordkeeping requirements for restricted-use pesticides. This rule would apply in those that do not.

The kind of information that must be kept is:

1. The branch or product name, formulation and EPA registration number of the restricted-use pesticide that was applied.

2. The total amount and rate of application of the restricted-use pesticide applied.

3. The address or location, the size of area treated, the target pest and the crop, commodity or stored product to which a restricted-use pesticide was applied.

4. The month, day, and year on which the application occurred.

5. The name, address and certification number (if applicable) of the certified applicator who made or who supervised the application.

The information must be recorded in a timely manner following pesticide application and records must be retained for two years after the date of the restricted-use pesticide application.

The commercial applicator must provide a copy of the records to the person for whom such an application was made within 30 days of the application.

Certified applicators must also maintain the records in a manner accessible by authorized representatives of the USDA or state pesticide regulators; make such records available to those officials upon oral request; and permit officials to copy the records.

This rule was issued by USDA. EPA has separate rules for recordkeeping of restricted-use applications.

Q & A Jay Feldman, national coordinator, National Coalition Against the Misuse of Pesticides.

Q. Your group believes cities and towns should be able to have their own regulations such as requirements to post signs of applications. But many states, in response to an industry push, are prohibiting those local rules. How is your effort to stop that going?

A. I wish we could say we could stop it in every state, but we can't. There are bright spots. In Maryland we've been able to literally stop the industry in its tracks in seeking pre-emption but it's an uphill battle.

Q. The industry argues that local governments should be pre-empted from having their own rules so regulations are more consistent. Why don't you agree?

A. The current standards that govern pesticides use, that is label restriction, are not adequate enough to ensure protection of human health on the ground on a day to day basis in the community. That's why it's essential that local governments be able to look at use patterns and sensitive sites and make determinations on appropriate use in that community.

Where there are sensitive sites in the community such as day care centers and school grounds it's really appropriate for the community to step in and look at use patterns and try to control use that's protective of the public.

Q. Is the certification and training system for applicators in this country adequate?

A. We've always thought the "under the supervision" clause (the part of FIFRA that says a restricted-use pesticide must be used under the direct supervision of a certified applicator) was enough of a loophole to allow for untrained people to be handling toxic materials in an effectively unsupervised manner. The idea of "under the supervision" is to allow for offsite supervision, which equates with meaningless supervision. That creates a loophole that results in untrained workers handling pesticides at harm to both themselves and to customers and clients.

In terms of training ... we're seeing an improvement in that area, but I don't believe the kind of training goes on that gives applicators a full appreciation for the fact they are handling very toxic materials. There's a lack of seriousness attached to the (pest control and lawn care) industry.

Training by itself is not the answer. Training plus enforcement and private right of action (citizens being able to sue) makes for a complete pesticide control system. So while agriculture departments continue to have jurisdiction over pesticides there continues to be minimal money invested in enforcement. There are relatively small penalties ... and no ability of citizens to enforce the law ... we will continue to see misuse situations.

Chapter 5

DEAD RATS AND YOU

Ultimately, the reason you will get to use a pesticide or not will come down to science. Often incomprehensible science, involving dead rats and complex mathematical formulas used to guess the risk from exposure to different chemicals. It may be useful to know what the scientific process is that pesticides go through and what terms such as "LD-50" mean.

HOW PESTICIDES ARE ASSESSED AT EPA

Remember that list of a couple dozen tests manufacturers must conduct on their products and submit to EPA as part of the federal registration process (it was in chapter 2)? Those studies all go into files that are reviewed by specialists in EPA called product managers.

These days, there aren't very many new chemicals being reviewed for the first time by these managers. But because all the old pesticides are being retested using newer scientific methods, there are huge numbers of studies being looked at within EPA. Also, additional studies may be required for pesticides that are on the market but seem to be causing problems and may need to be suspended. Among the things the EPA scientists look at is if "good laboratory practices" have been used by the lab that conducted the study. That means the lab was clean, the intruments calibrated and good records kept. The agency also scrutinizes whether the study represents "good science," by being conducted carefully and according to the standards of the time.

Unfortunately, EPA is now rejecting a surprisingly large percentage of studies that are being submitted for reregistration because they don't meet those standards. Another problem is when a laboratory that does a lot of work for industry is found to be faking the results. There have been two such scandals, resulting in the invalidation of large numbers of studies.

EPA's scientific decisions are reviewed by a panel of outside experts called the Scientific Advisory Panel. This board meets periodically to look at EPA decisions such as canceling a pesticide's use, conducting a special review of a problem pesticide or establishing scientific policies on carcinogen rankings or risk assessment.

Although the SAP's findings are only advisory, they often carry a lot of political weight when top EPA officials are deciding whether a pesticide should stay on the market.

It is during these meetings that the uncertainty and politics within science become clear. People tend to assume that science is black and white, and above political wrangling. But like anything else, scientists looking at the same evidence can have very different points of view about it, particularly if they work for the industry or an interest group.

Often, the SAP is asked to decide whether EPA has classified a pesticide correctly on its cancer assessment guidelines. Those guidelines rank pesticides from A, human carcinogen potent, to E, where there is evidence it is not a carcinogen. Most arguments are over whether a pesticide is considered a "probable" or "possible" carcinogen (a B or a C). That distinction can be important, especially when environmental groups issue lists of pesticides considered to be carcinogenic. Often they'll use the ones listed by EPA as "probable" carcinogens. Because of the difficulty of pinning down what causes cancer, there are only a handful of substances that EPA has categorized as definite cancer-causers.

Also, since 1988 EPA scientists and policymakers have been working to revise the cancer risk assessment guidelines to reflect advances in understanding about how chemicals cause cancer.

Sometimes the debate rages over whether EPA should use studies that show tumors in mice as indicative of what the pesticide would do in humans. Also, there is criticism of the fact that the mice or other study animals are often force-fed unnaturally large amounts of a substance. But federal scientists are convinced that the current system works and that their conservative assumptions about carcinogens best protect public health, which is their mission.

Many people are urging EPA to look at health effects other than cancer, such as neurotoxic damage. That refers to the central nervous system and can involve temporary symptoms such as nausea or permanent nerve or brain damage.

The problem with testing chemicals for neurotoxicity is that most of the symptoms are difficult to find — how does a rat or a dog tell a scientist it feels bad? Still, certain types of chemicals known to cause these problems in people are already required to be tested, and EPA plans to require more chemicals to undergo neurotoxicity tests.

Not only do the scientists look at effects of chemicals on human beings but they also look at pesticides in the environment. For instance, the agency wants to know if a herbicide used on large expanses of lawn is likely to be toxic to birds that might land on the grass. However, agency officials have said recently they may need to pull away from requiring so many environmental studies because they are expensive, may not be necessary for some chemicals and take two or three years to conduct, thereby delaying the already lengthy process for approving a pesticide's use.

RISK ASSESSMENT

Once the biological studies have been examined, the statisticians take over in a process known as risk assessment. In the modern sense, the results of the tests are thrown into mathematical models and the outcome is often a number such as 1 in 1 million. Under EPA's system, that would mean that of a population of 1 million people all exposed to a substance 24 hours a day, over their entire lifetimes, one person would likely contract cancer who would not have been expected to without the exposure to that substance.

Risk assessment is supposed to give people a way to get a definitive grip on whether an exposure is too risky. It is routinely used by EPA and lawmakers to decide the cutoff point at which people should no longer be exposed to a particular substance.

The problem with that, many of the scientists say, is that risk assessment is a highly theoretical proposition. For instance, they don't really mean that one person in a million is going to get cancer. They don't know that. And that result becomes even more theoretical when you start considering all the assumptions that went into it: That someone would be exposed their whole life, all the time, at the maximum amount, and that the person would also eat food contaminated with the substance, and that they would fit the height and weight assumptions in the model. Plus, the risk number is multiplied 1,000-fold as a safety precaution. So the final numbers are not supposed to be taken literally, the scientists say.

But politicians and regulators need something to go on, they argue, and risk assessment numbers are the most tangible thing they've got. That's why debates over the safety of Alar on apples, for instance, got so complex: Some said it is wrong to warn the public of risk if the science is so uncertain. But others said the government has a responsibility to the public to tell them if there's any risk at all.

There is also a school of thought that says EPA risk assessment methods may actually underestimate the risk of pesticides in food on children and the elderly, because the model assumes the impact of a chemical on a healthy adult.

Also impacting how the public sees risk is the fact that most Americans appear to believe they should not be exposed to any undue risk at all. This is despite the fact that smoking or driving a car are many thousands of times riskier than eating food with pesticide residues. The difference, risk managers say, is that people know the risk of smoking and consciously accept it. What they don't like is having someone else force a risk on them that they didn't choose or didn't know about.

Those thought processes explain why scientists and the public come up with quite different conclusions when asked about their biggest environmental worries.

For instance, a March 1990 Roper Poll, reported in Science magazine, showed that science advisers to the EPA put global climate change and ozone depletion as the most serious environmental problems. But the public ranked hazardous waste sites and water pollution from industrial wastes as the top concerns.

About half those polled listed pesticides in food as a serious problem, while "application of pesticides" was among the scientists' top 11 problems.

Another part of EPA's assessment of whether a pesticide should be used is the risk-benefit balancing it must do. Unlike some other environmental laws, the federal pesticides statute says that EPA must allow a pesticide on the market if its benefits outweigh its risks.

Pesticide makers and users have long complained that the "benefits" side of the equation is missing realistic information about how much pesticides are actually being used out there. EPA always assumes pesticides are being used at their maximum label rates. By plugging in more realistic use rates, some pesticides might actually be seen as safer than they are now.

The only way to remedy that, many say, is by requiring pesticide users to keep records of their applications, which many users, especially farm groups, have fought. But that may change eventually. It's already being done in California, whose state government also does its own scientific analysis and risk assessment of pesticides used within its borders. It is just beginning to compile the results of those pesticide use reports for use in risk assessment.

A BRIEF EXPLANATION OF TOXICOLOGY

Toxicology — the science that studies the harmful effects of chemicals — is the discipline that most affects pesticide users.

The following is a brief description of some of the basics of toxicology. Much of it is gleaned from a book called "The Dose Makes the Poison" by Alice Ottoboni and "Elements of Toxicology and Chemical Risk Assessment" by Environ Corp.

To begin with, chemicals are everywhere. All matter is composed of chemical elements. Chemicals fall into two major categories: Natural (made by natural processes) and synthetic (made by human beings). Many chemicals fall into both categories.

Many people assume natural chemicals are better than synthetic, but Ottoboni, a retired toxiologist for California's Department of Health Services, argues there's no evidence of that. "The fact is that a living organism cannot distinguish chemicals by their origin, be they from nature or the laboratory. It can only distinguish between chemicals it can use to make more of itself and chemicals it cannot use."

The public often views all pesticides as being toxic chemicals, she notes. The fact is that many pesticides are no more toxic than many non-pesticide chemicals that we encounter in our daily lives, Ottoboni writes.

Of course, environmentalists would argue that pesticides are created specifically to kill things, so the likelihood they would be toxic is greater than the average chemical.

Ottoboni notes that the toxicity of a chemical is dependent upon its use. For instance, boric acid is regulated as a drug when used as an antiseptic eye wash, as a household product when used in laundry detergents, as an insecticide when used to kill roaches, as a herbicide when used to kill weeds and as a flame retardant when used to fireproof fabrics.

Ottoboni's main point is that a chemical's effects are entirely dependent upon the dose involved and how often the exposure is. That is known as the dose-time relationship. "It is impossible to study every possible combination of dose and time, so toxicologists study effects that occur at the two extremes and then make judgements of what might happen in between," she writes.

Also important is the "route of exposure," or how the chemical enters the body: orally, or by mouth; inhalation, or through the lungs; dermally, or through the skin. A chemical's effects can be very different depending on the route of exposure.

Other factors that influence a chemical's toxicity to humans or animals include the species; its age and sex; nutrition, state of health, individual susceptibility, chemical combinations (known as synergism) and adaptation (when repeated exposures to small quantities of a chemical makes a person tolerant to larger exposures).

There are two different types of toxicity that are important to distinguish between when talking about pesticide exposure. The first is acute toxicity, or poisoning from a brief, single exposure to a substance. A lot is known about acute toxicity both because it is relatively easy to study in the laboratory and because information is available from accidental exposures and suicides.

A "Lethal Dose 50" test is the way scientists find out the acute oral toxicity of chemicals. That means giving several groups of laboratory animals increasing doses of the substance. Ideally, Ottoboni explains, the smallest dose will kill none of the animals and the largest will kill all of them. The animals are observed for 15 days. The LD-50 refers to the dose that would kill half the animals.

Similar tests are used to find out the acute toxicity for dermal and inhalation routes. For this information, the chemical is put on the animal's skin or is breathed.

The other major category of toxicity is called chronic toxicity, which means the ability of a chemical to cause harm from repeated exposures over long periods of time. Testing involves long-term or lifetime exposure of groups of animals to levels of chemicals.

EPA requires both acute (one-time) and chronic (long-term) toxicity tests for all pesticides it regulates, as well as reproduction, mutagenicity (mutations) and carcinogenicity tests.

Some Terms

Acute toxicity: The adverse effects from a one-time exposure.

Cancer: An abnormal, potentially unlimited, new tissue growth.

Carcinogen: A substance that increases the incidence of cancer.

Chronic toxicity: The adverse effects from many repeated exposures.

Dermal: Related to the skin.

Dermatitis: Inflammation of the skin.

DNA: Deoxyribonucleic acid; the complex biochemical in the nucleus of cells that carries the genetic code (the code that tells the cell how to reproduce itself exactly).

Epidemiology: The study of the distribution and causes of diseases and injuries in human populations.

Exposure: The level of exposure to substance in the vicinity of a portal of entry to the body that may be available for absorption.

Genotoxic: Able to interact with and damage the genetic material, DNA of the cell, often causing a mutation. Some mutations may cause cancer to develop.

Ingestion: Intake of a substance through the mouth.

Inhalation: Intake of a substance through the lungs.

LD 50: Lethal dose, 50 percent; the dose of a chemical that is lethal to 50 percent of those exposed. Usually refers to the number of laboratory animals affected by a test.

Metabolism: Chemical change in a substance occurring in the body.

Mutation: A heritable change in the structure or sequence of the DNA that carries the "blueprint" for the normal functioning of the cell and which changes the function or behavior of the cell.

Natural chemical: A chemical made by natural processes.

Pesticide: A substance used to kill something that has been determined to be a pest.

Risk: The nature and probability of occurrence of an unwanted, adverse effect on human life or health, or on the environment.

Risk assessment: The characterization of the potential adverse effects on human life or health or on the environment.

Subchronic: Intermediate in duration between acute and chronic.

Synergism: Interaction between two substances that results in a greater effect than both of the substances could have produced had they acted independently.

Synthetic chemical: A chemical made in the laboratory.

Teratogenic: Causing malformations or birth defects as a result of exposure during embryonic or fetal development.

Toxic effect: Any change in an organism that results in impairment of functional capacity of the organism (as determined by anatomical, physiological, biochemical or behavioral parameters); causes decrements in the organism's ability to maintain its normal funtion; or enhances the susceptibility of the organism to the deleterious effects of other environmental influences.

Toxicology: The science that studies the harmful systemic effects of chemicals.

Tumor: Literally, a swelling. A growth of new tissue.

Q & A James M. Seiber, Ph.D., pesticides researcher from the University of Nevada, Reno, who specializes in environmental toxicology — how pesticides act in the environment and expose people. He worked out of UC Davis for many years as associate dean of research.

Q. Do current regulations adequately protect people who apply pesticides?

A. In general, exposure to people working out in the fields is greatest. Residues in food are really a drop in the bucket. It's the migrants, the applicators who are really getting the exposures. Good training is critical, but we've been very lackadaisical in the past about it. The rules are there, but there isn't enough enforcement. When workers are out in the field there's nobody there to remind them.

Q. But much of the public attention has been on residues in food, and the public has pushed politicians in that direction.

A. That interest is misplaced. The real problem is with workers. We haven't done enough there. There's been resistance ... with small business, it's one more burden they have to bear. We need simpler ways of training people.

Q. A major problem seems to be getting applicators to wear their protective equipment, such as goggles and gloves.

A. It's clumsy, it gets in the way. The equipment is not well designed. They're using stuff that was invented for miners and it's being used in the fields. Somebody needs to redesign it so it's easier to use.

Q. What about product labels — are they comprehensible?

A. The label is so lengthy it becomes like a technical document. It's so long and incomprehensible. There needs to be a shortened version of the label, using warning sings. That boils it down to a simple skull and crossbones.

Q. Are manufacturers doing an adequate job testing their products for acute exposures that workers are likely to get?

A. Yeah, I think they're dong all kinds of tests in dermal and inhalation. They do a lot better job of the physical, chemical characterization of a product, when it will break down, what it does in water. In general, products tends to be safer than they were. Everything's going in the right direction. It's getting the information to the person who's using it that's the problem.

Q. Farmworker groups in California are trying to get more scientific attention to health problems of workers and their families from pesticide exposure. Is there anything to it?

A. Yeah, I think that's an issue. Farmworkers bring a lot of residues home on their clothing. They get a pretty high exposure, as do their families. I don't think we really understand that. I worked on one of those studies (of an alleged cancer cluster in California's Central Valley) and we just didn't have the information.

Q. The scientific community has trouble agreeing about when a pesticide is safe. There always seems to be a new kind of health effects test. Is it possible to declare a pesticide safe to use once and for all?

A. As painful as reregistration (the re-testing of older pesticides to meet updated standards) has been, I think it's been a good thing. You've gotten these products to be tested more. It's getting pretty complete. So complete, in fact, that companies can't afford to do all the testing, so they drop products. That's the bad side of it. The other danger is we won't have new promising products come out in the future.

Q. It seems much of the pesticide-using industry is trying to avoid all that by just using fewer or no pesticides.

A. Yes, and that's good. Minimization is a better term. Pick out the safer methods. Biocontrol and all those derivatives are starting to perk up and become competitive. They're getting more reliable.

Chapter 6

LABELS

Because the label of a pesticide is the official, legal direction for how to use it, both the contents on that label and understanding them thoroughly are vital to the pesticide applicator.

For professional use pesticides — as opposed to home-and-garden pesticides used by consumers — the word label can sound misleading. For someone used to reading the basic information and warnings that would fit on the outside of a container, the pesticide labels that accompany many products meant for farmers and professionals to use can be daunting. Some are actually booklets, many pages long.

The EPA has specific rules for what kind of information must show up on the label. That includes all the names the pesticide goes by, both trade and common names, along with information such as its toxicity category, its specific ingredients, how it should be used, how it shouldn't be used, how it should be stored and disposed of and what to do if someone is poisoned by it.

Labels can become quite complex and change on a regular basis because the Environmental Protection Agency is regularly issuing new restrictions or warnings or creates new programs that must be explained on the labels.

Constantly changing label content has become a bone of contention between the agency and pesticide manufacturers, who complain that it costs them a lot of money to print up all new labels for their products on a regular basis. EPA usually gives them a certain amount of time to get the products with the old labels off the market.

There have also been concerns for many years that pesticide labels have so much information on them that they are to confusing for users, who aren't likely to take the time to read the whole thing out there in the field or in a building to be sprayed. Compounding the problem is the fact that many applicators don't read English and that the products are often sold in other countries where applicators may read another language or not at all.

Those concerns led to an effort in the mid-1980s to create new, clearer labels with things such as color-coding, so someone would immediately know how toxic the product was, and pictographs rather than words to bridge the language barrier. There was also an effort started by a pesticide manufacturers group to substantially shorten each pesticide's label by taking a lot of the generic information placed on all labels and putting it into a single manual that every applicator would keep with him or her.

While there was a lot of talk about these things a few years ago, they were soon put on the back burner at EPA because of other pressing matters, such as reregistering hundreds of old pesticides. Agency officials now say they doubt there is money available in government coffers to pursue the label improvement project.

However, there is a new move afoot within EPA to improve the way its employees handle changes to labels, which are sometimes initiated by the agency and sometimes by the manufacturer. Because those change requests are reviewed by different people in EPA, called product managers, there can be inconsistencies in what language is allowed and what is not. That can lead to confusion for users, so EPA officials are looking at how they can simplify the process.

HOW TO READ A LABEL

It may seem obvious how to read a label, but it's not necessarily that easy. For instance, the words "danger," "warning" and "caution" may seem quite similar but there are subtle differences in what they mean on a pesticide label. Included here is a typical pesticide label, for atrazine, one of the most commonly used pesticides in the United States. The following is some general advice on reading labels along with the label itself, so you can see how it's actually done. Much of the following advice was taken from a well-written basic training manual issued by the Colorado Department of Agriculture.

Here are some of the things that must be included on a label, by the federal pesticides law (also remember that "the label is the law" — doing anything contrary to a product's label is illegal):

—Brand name, common name and chemical name: The brand name, or trade name, is the manufacturer's proprietary name for a product. The common name is the generic word for the active ingredient inside. The chemical name is the specific, scientific name for the chemical, and is often long and difficult to pronounce.

—Use classification: Pesticides are either for general use, by anyone, or for restricted use. Restricted-use pesticides can only be used by a certified applicator or someone under a certified applicator's direct supervision, according to FIFRA.

—Ingredients: The active ingredient is the chemical that actually does the job the product is intended to do. But the bulk of the pesticide is usually inert ingredients, such as wetting agents, solvents, carriers or fillers. Inert ingredients do not have to be named because manufacturers often believe they are proprietary business information. However, that does not mean they are necessarily harmless, and EPA has a program going to study potential health effects from inert pesticide ingredients.

—Use of the pesticide: The labels lists only the uses for which the pesticide is registered, which are the only legal uses for the product. So if a product is registered only for use on avocadoes in California, it can't be used on wheat in Indiana. The purpose for which you intend to use a particular pesticide product must appear on the label when you buy the product.

—Directions for use: These are specific directions for using the pesticide properly and tell you what time of year, weather conditions, type of crop and how much pesticide to use. The directions state how many applications can be made within a given period of time and how much time to wait until going back into the field or building after application. Remember that the rate on the label isn't just a suggestion for how the product works best, it's a legal limit which, if exceeded, can get an applicator in trouble.

—Safety information, signal words and precautions: Certain signal words are required on every label. These words are "danger-poison" and the skull and crossbones (all in red) or "warning" or "caution," depending on the hazard of the particular product to the user. The label must also carry the statement "keep out of the reach of children." The signal word required depends on the toxicity and potential hazard of the active ingredient and also the particular formulation, or amount of active ingredient, in the product.

"Danger-poison" and the skull and crossbones are required on labels of all highly toxic compounds, which are those whose acute oral lethal dose-50 of 5 to 50 milligrams per kilogram of body weight. In other words, in tests done on laboratory animals, half died after being fed between 5 mg and 50 mg of the active ingredient as a ratio of the animal's body weight.

"Warning" is required on moderately toxic compounds, those having an LD50 of between 50 and 500 mg/kg.

"Caution" is required on labels for all slightly toxic compounds (with LD 50s of 500 to 5,000 mg/kg) and relatively non-toxic compounds (LD 50 of 5,000 mg/kg and above).

To put in perspective what those toxicity hazards mean, a highly dangerous pesticide could kill a man with a taste to a teaspoon of the pesticide. A moderately hazardous product would take a teaspoon to a tablespoon, while slightly hazardous ones would take an ounce to a pint.

The pesticide label also gives detailed emergency first-aid measures. The pesticide label is the most important information that can be provided to a physician if a person is suspected of being poisoned.

—EPA registration number and establishment number: The EPA registration number indicates the product has been registered with the EPA, as required by law. The establishment number is a special number assigned to the plant that manufactured the pesticide for the company.

—Name and address of manufacturer or registrant: This is the name and address of the company manufacturing the product and offering it for sale.

The Colorado handbook also offers some good advice about the five times someone should read the pesticide label.

The first time to read the label is when you buy the pesticide to make sure it's the best chemical for the purpose and to be sure you have the proper equipment to apply it.

The second time is before you mix the pesticide to make sure you have any protective equipment required, know what it can be mixed with and how to mix it.

The third time is before you apply the pesticide to determine any safety measures to follow, how and how much to apply it and any special instructions. Check again before going back into the field or building to be sure you've followed directions on any waiting period.

The fourth time is before you store the pesticide to find out where and how to store it and what it should not be stored with. The fifth time is before you dispose of the excess pesticide in the container to determine how and where to dispose of the pesticide, how to clean a pesticide container and how to dispose of the container.

Table 2

RESTRICTED USE PESTICIDE
(GROUND AND SURFACE WATER CONCERNS)

FOR RETAIL SALE TO AND USE ONLY BY CERTIFIED APPLICATORS OR PERSONS UNDER THEIR DIRECT SUPERVISION AND ONLY FOR THOSE USES COVERED BY THE CERTIFIED APPLICATOR'S CERTIFICATION.

THIS PRODUCT IS A RESTRICTED-USE HERBICIDE DUE TO GROUND AND SURFACE WATER CONCERNS. USERS MUST READ AND FOLLOW ALL PRECAUTIONARY STATEMENTS AND INSTRUCTIONS FOR USE IN ORDER TO MINIMIZE POTENTIAL FOR ATRAZINE TO REACH GROUND AND SURFACE WATER.

CIBA–GEIGY

AAtrex® Nine-O®

HERBICIDE

For season-long weed control in corn, sorghum and certain other crops

Active Ingredients: Atrazine: 2-chloro-4-ethylamino-6-isopropylamino-s-triazine 85.5%
Related compounds 4.5%
Inert Ingredients: 10.0%
Total: 100.0%

AAtrex Nine-O is a water dispersible granule.

EPA Reg. No. 100-585   EPA Est. 100-LA-1

KEEP OUT OF REACH OF CHILDREN.

CAUTION

See additional precautionary statements and directions for use inside booklet.   CGA 7L101M 052

DIRECTIONS FOR USE AND CONDITIONS OF SALE AND WARRANTY

IMPORTANT: Read the entire Directions for Use and the Conditions of Sale and Warranty before using this product. If terms are not acceptable, return the unopened product container at once.

Conditions of Sale and Warranty

The Directions for Use of this product reflect the opinion of experts based on field use and tests. The directions are believed to be reliable and should be followed carefully. However, it is impossible to eliminate all risks inherently associated with use of this product. Crop injury, ineffectiveness, or other unintended consequences may result because of such factors as weather conditions, presence of other materials, or the manner of use or application all of which are beyond the control of CIBA–GEIGY or the Seller. All such risks shall be assumed by the Buyer.

CIBA–GEIGY warrants that this product conforms to the chemical description on the label and is reasonably fit for the purposes referred to in the Directions for Use subject to the inherent risks referred to above. **CIBA–GEIGY makes no other express or implied warranty of Fitness or Merchantability or any other express or implied warranty. In no case shall CIBA–GEIGY or the Seller be liable for consequential, special, or indirect damages resulting from the use or handling of this product.** CIBA–GEIGY and the Seller offer this product, and the Buyer and user accept it, subject to the foregoing Conditions of Sale and Warranty, which may be varied only by agreement in writing signed by a duly authorized representative of CIBA–GEIGY.

DIRECTIONS FOR USE

It is a violation of federal law to use this product in a manner inconsistent with its labeling.

FAILURE TO FOLLOW THE DIRECTIONS FOR USE AND PRECAUTIONS ON THIS LABEL MAY RESULT IN POOR WEED CONTROL, CROP INJURY, OR ILLEGAL RESIDUES.

Do not apply this product in such a manner as to directly or through drift expose workers or other persons, except those knowingly involved in the application. The area being treated must be vacated by unprotected persons.

Do not enter treated areas without protective clothing until sprays have dried.

Because certain states may require more restrictive reentry intervals for various crops treated with this product, consult your State Department of Agriculture for further information.

Written or oral warnings must be given to workers who are expected to be in a treated area or in an area about to be treated with this product. Oral warnings must be given which inform workers of areas or fields that may not be entered without specific protective clothing until sprays have dried, and appropriate actions to take in case of accidental exposure as described under **Precautionary Statements** on this label. When oral warnings are given, warnings shall be given in a language customarily understood by workers. Oral warnings must be given if there is reason to believe that written warnings cannot be understood by workers. Written warnings must include the following information: "CAUTION. Area treated with AAtrex Nine-O on (date of application). Do not enter without appropriate protective clothing until sprays have dried. In case of accidental exposure, flush eyes or skin with plenty of water. Call a physician if irritation persists. Remove and wash contaminated clothing before reuse."

Do not apply this product through any type of irrigation system.

General Information

This herbicide controls many annual broadleaf and grass weeds in corn, sorghum, sugarcane, and certain other crops specified on this label. This product may be applied before or after weeds emerge.

Following many years of continuous use of this product and chemically related products, biotypes of some of the weeds listed on this label have been reported which cannot be effectively controlled by this and related herbicides. Where this is known or suspected, and weeds controlled by this product are expected to be present along with resistant biotypes, we recommend the use of this product in combinations or in sequence with other registered herbicides which are not triazines. If only resistant biotypes are expected to be present, use a registered non-triazine herbicide. Consult with your state Agricultural Extension Service for specific recommendations.

Since this product acts mainly through root absorption, its effectiveness depends on moisture to move it into the root zone. If weeds develop, a shallow cultivation or rotary hoeing will generally result in better weed control.

This product is noncorrosive to equipment and metal surfaces, nonflammable, and has low electrical conductivity.

Avoid using near adjacent desirable plants or in greenhouses, or injury may occur.

To avoid spray drift, do not apply under windy conditions. Avoid spray overlap, as crop injury may result.

Where the use directions give a range of rates, use the lower rate on coarse-textured soil and soil low in organic matter; use the higher rate on fine-textured soil and soil high in organic matter.

Note: CIBA–GEIGY does not recommend applications in combination with other herbicides or oils, except as specifically described on the label or in literature published by CIBA–GEIGY.

Application Procedures

Ground application: Use conventional ground sprayers equipped with nozzles that provide accurate and uniform application. Be certain that nozzles are uniformly spaced and are the same size. Calibrate sprayer before use and recalibrate at the start of each season and when changing carriers. Unless otherwise specified, use a minimum of 10 gals. of spray mixture/A for all preplant incorporated, preplant surface, preemergence, and postemergence applications (with or without oil or surfactant) with ground equipment.

Use a pump with capacity to (1) maintain 35-40 psi at nozzles, (2) provide sufficient agitation in tank to keep mixture in suspension, and (3) to provide a minimum of 20% bypass at all times. Use centrifugal pumps which provide propeller shear action for dispersing and mixing this product. The pump should provide a minimum of 10 gals./minute/100 gal. tank size circulated through a correctly positioned sparger tube or jets.

Use screens to protect the pump and to prevent nozzles from clogging. Screens placed on the suction side of the pump should be 16-mesh or coarser. Do not place a screen in the recirculation line. Use 50-mesh or coarser screens between the pump and boom, and where required, at the nozzles. Check nozzle manufacturer's recommendations.

For band applications, calculate amount to be applied per acre as follows:

$$\frac{\text{band width in inches}}{\text{row width in inches}} \times \frac{\text{broadcast rate}}{\text{per acre}} = \frac{\text{amount needed}}{\text{per acre of field}}$$

CGA 130-560Q

Table 2

AAtrex® Nine-O®

Aerial application: Use aerial application only where broadcast applications are specified. Apply in a minimum of 1 gal. of water for each 1 lb. of AAtrex Nine-O applied per acre. For postemergence treatments on corn and sorghum, apply recommended rate in a minimum of 2 gals. of water/A. Avoid applications under conditions where uniform coverage cannot be obtained or where excessive spray drift may occur.

Avoid application to humans or animals. Flagmen and loaders should avoid inhalation of spray mist and prolonged contact with skin, and should wash thoroughly before eating and at the end of each day's operation.

Application in water or liquid fertilizer: Nitrogen solution or complete liquid fertilizer may replace all or part of the water as a carrier for preemergence, preplant incorporated, or preplant surface ground application on corn and sorghum. Check the compatibility of this product with liquid fertilizer and/or nitrogen solution as shown below before use. Do not apply in nitrogen solution or complete liquid fertilizer after corn or sorghum emerges or crop injury may occur.

Compatibility Test: Since liquid fertilizers can vary, even within the same analysis, always check compatibility with herbicide(s) **each time before use.** Be especially careful when using **complete** suspension or fluid fertilizers as serious compatibility problems are more likely to occur. Commercial application equipment may improve compatibility in some instances. The following test assumes a spray volume of 25 gals. per acre. For other spray volumes, make appropriate changes in the ingredients. Check compatibility using this procedure.

1. Add 1 pint of fertilizer to each of 2 one-quart jars with tight lids.

2. To one of the jars add ¼ tsp. or 1.2 milliliters of a compatibility agent approved for this use, such as Compex* or Unite* (¼ tsp. is equivalent to 2 pts. per 100 gals. spray). Shake or stir gently to mix.

3. To **both** jars add the appropriate amount of herbicide(s). If more than one herbicide is used, add them separately with dry herbicides first, flowables next, and emulsifiable concentrates last. After each addition, shake or stir gently to thoroughly mix. The appropriate amount of herbicides for this test follows.

   Dry herbicides: For each pound to be applied per acre, add 1.5 level teaspoons to each jar.

   Liquid herbicides: For each pint to be applied per acre, add 0.5 teaspoon or 2.5 milliliters to each jar.

4. After adding all ingredients, put lids on and tighten, and invert each jar ten times to mix. Let the mixtures stand 15 minutes and then look for separation, large flakes, precipitates, gels, heavy oily film on the jar, or other signs of incompatibility. Determine if the compatibility agent is needed in the spray mixture by comparing the two jars. If either mixture separates, but can be remixed readily, the mixture can be sprayed as long as good agitation is used. If the mixtures are incompatible, test the following methods of improving compatibility: (A) slurry the dry herbicide(s) in water before addition, or (B) add ½ of the compatibility agent to the fertilizer and the other ½ to the emulsifiable concentrate or flowable herbicide before addition to the mixture. If incompatibility is still observed, do not use the mixture.

Application in water plus emulsifiable oil or oil concentrate: Adding emulsifiable oil (petroleum-derived, petroleum-derived oil concentrate, or single or mixed crop-derived oil concentrate) to postemergence water-based sprays in corn and sorghum may improve weed control. However, under certain conditions, the use of either type of oil may seriously injure the crop. To minimize this possibility, observe the following directions:

Use one of the following properly emulsified:

1. A suitable oil concentrate containing at least 1% but not more than 20% suitable emulsifier or surfactant blend.

2. Petroleum-derived oil containing at least 1% suitable emulsifier.

Note: In the event of a compatibility problem when mixing oil with AAtrex Nine-O and water, a compatibility agent such as Compex or Unite should be used. Any of the above oils contaminated with water or other materials can cause compatibility problems and/or crop injury.

Mixing procedures – all uses: (1) Be sure sprayer is clean and not contaminated with any other materials, or crop injury or sprayer clogging may result. (2) Fill tank ¼ full with clean water, nitrogen solution, or complete liquid fertilizer. (3) Start agitation. (4) Be certain that the agitation system is working properly and creates a rippling or rolling action on the liquid surface. (5) Pour product directly from bag into tank. (6) Continue filling tank until 90% full. Increase agitation if necessary to maintain surface action. (7) Add emulsifiable oil, oil concentrate, or tank mix herbicide(s) after this product is thoroughly suspended. (8) Finish filling tank. (9) Empty tank as completely as possible before refilling to prevent buildup of oil or emulsifiable concentrate residue. Maintain agitation to avoid separation of materials. (10) If an oil or emulsifiable concentrate film starts to build up in tank, drain it and clean with strong detergent solution or solvent. (11) Clean sprayer thoroughly immediately after use by flushing system with water containing a detergent.

Rotational Crops – All Uses: (1) Do not rotate to any crop except corn or sorghum until the following year, or injury may occur. (2) If applied after June 10, do not rotate with crops other than corn or sorghum the next year, or crop injury may occur. (3) In the High Plains and Intermountain areas of the West where rainfall is sparse and erratic or where irrigation is required, use only when corn or sorghum is to follow corn or sorghum or when a crop of untreated corn or sorghum is to precede other rotational crops. (4) In eastern parts of the Dakotas, KS, western MN, and NE, do not rotate to soybeans if the rate applied to corn or sorghum was more than 2.2 lbs./A or equivalent band application rate, or soybean injury may occur. (5) Injury may occur to soybeans planted the year following application on soils having a calcareous surface layer. (6) Do not plant sugar beets, tobacco, vegetables (including dry beans), spring-seeded small grains, or small-seeded legumes and grasses the year following application, or injury may occur.

AAtrex Nine-O Applied Alone – Corn or Grain Sorghum*

Preplant Surface-Applied, Preplant Incorporated, or Preemergence (or Postemergence at 2.2 lbs./A With Oil)

Broadleaf and Grass Weeds Controlled

barnyardgrass (watergrass)***	cocklebur**	nightshade
giant foxtail**	groundcherry	pigweed
green foxtail***	jimsonweed	purslane
large (hairy) crabgrass**	kochia	ragweed
wild oats	lambsquarters	sicklepod**
witchgrass	annual	velvetleaf
(Panicum capillare)***	morningglory	(button-
yellow foxtail***	mustards	weed)***

Postemergence with Emulsifiable Oil or Oil Concentrate in Water (at 1.3 lbs./A)

Broadleaf Weeds Controlled

annual morningglory	lambsquarters	ragweed	wild buckwheat
cocklebur	mustards	smartweed	velvetleaf**
jimsonweed	pigweed		

*Where there are state/local requirements regarding atrazine use (including lower maximum rates and/or greater setbacks) which are different from the label, the more restrictive/protective requirements must be followed.

Certain states may have established rate limitations within specific geographical areas. Consult your state lead pesticide control agency for additional information. It is a violation of this label to deviate from state use regulations.

** Partial control only.

*** Partial control only on medium and fine-textured soils.

Corn

Preplant Surface-Applied (Broadleaf and grass control): Use on medium- and fine-textured soil with minimum-tillage or no-tillage systems only in CO, IA, IL, IN, KS, KY, MN, MO, MT, ND, NE, SD, WI, and WY. Apply the recommended rate of AAtrex Nine-O shown in Table 1 up to 45 days prior to planting. On **coarse-textured** soils, do not apply more than two weeks prior to planting. If an unsatisfactory length of weed control results from adverse environmental conditions following early treatment, a follow-up application of an appropriately labeled herbicide may be used. If the follow-up treatment includes atrazine, do not exceed the labeled rate for corn indicated in Table 1.

If weeds are present at the time of treatment, apply in tank mix combination with a contact herbicide (for example, Gramoxone® Extra or Roundup®). Observe directions for use, precautions, and restrictions on the label of the contact herbicide.

Note: To the extent possible, do not move treated soil out of the row or move untreated soil to the surface during planting, or weed control will be diminished.

Preplant Incorporated (Broadleaf and grass control): Broadcast in spring after plowing at rate in Table 1. Apply to the soil and incorporate before, during, or after final seedbed preparation. Avoid deep incorporation. For best results, apply within two weeks prior to planting.

Preemergence or At-Planting (Broadleaf and grass control): Apply during or shortly after planting before weed emergence, at rate in Table 1.

Postemergence (Broadleaf and grass control): Apply before weeds exceed 1.5 inches in height and before corn exceeds 12 inches in height at rates recommended in Table 1.

Table 1: Broadleaf and Grass Weed Control in Corn*

FOR ALL SOIL APPLICATIONS PRIOR TO CROP EMERGENCE

- **On Highly Erodible Soils (as defined by SCS)**

  If conservation tillage is practiced, leaving at least 30% of the soil covered with plant residues at planting, apply a maximum of 2.2 lbs./A as a broadcast spray.

  If the soil coverage with plant residue is less than 30% at planting, a maximum of 1.8 lbs./A may be applied.

- **On Soils Not Highly Erodible**

  Apply 2.2 lbs./A as a broadcast spray.

FOR POSTEMERGENCE APPLICATION

If no atrazine was applied prior to corn emergence, apply a maximum of 2.2 lbs./A broadcast. If a postemergence treatment is required following an earlier herbicide application, the total atrazine applied may not exceed 2.5 lbs. active ingredient (2.8 lbs. of this product) per acre per calendar year.

*Broadleaf control (eastern CO, western KS, western NE, NM, OK Panhandle, west TX, and eastern WY): On sand, loamy sand, sandy loam, mild to strongly alkaline soil, and all recently leveled soil, apply no more than 1.3 lbs./A, either preplant surface, preplant incorporated, or preemergence. On other soils in these areas, apply rate in Table 1 for broadleaf and grass control.

Where there are state/local requirements regarding atrazine use (including lower maximum rates and/or greater setbacks) which are different from the label, the more restrictive/protective requirements must be followed. Certain states may have established rate limitations within specific geographical areas. Consult your state lead pesticide control agency for additional information. It is a violation of this label to deviate from state use regulations.

Table 2

AAtrex® Nine-O®

Postemergence with emulsifiable oil or oil concentrate in water: Add the following volume of one of the type oils indicated for aerial or ground application unless the oil label specifies otherwise.

Type Oil	Ground Application	Aerial Application
Oil Concentrate (Crop or Petroleum-derived)	1 qt./A	½ - 1 qt./A
Petroleum-derived oil	1 gal./A	2 qts./A

Note: Crop-derived or petroleum-derived oil concentrates should contain at least 1% but not more than 20% suitable emulsifier or surfactant blend. Petroleum-derived oils should contain at least 1% suitable emulsifier.

Broadleaf and grass control: For postemergence control of those weeds listed under **Preplant Incorporated and Preemergence**, broadcast 2.2 lbs./A plus emulsifiable oil or oil concentrate after weed emergence, but before weeds reach 1.5 inches in height and before corn exceeds 12 inches in height.

Broadleaf control: For postemergence control of those weeds listed under **Postemergence with emulsifiable oil or oil concentrate in water,** broadcast 1.3 lbs./A plus emulsifiable oil or oil concentrate before pigweed and lambsquarters reach 6 inches in height and before all other weeds reach 4 inches in height. A cultivation may be necessary if all weeds are not controlled or if weeds regrow.

Precautions: For applications with emulsifiable oil or oil concentrate in water: (1) Inbred lines or any breeding stock may be severely injured by applications with emulsifiable oil or oil concentrate. (2) Adding other insecticides, herbicides, liquid fertilizers, or other materials is not recommended, because they may cause compatibility problems or crop injury. (3) Store and handle emulsifiable oil and oil concentrate carefully. Oil contaminated with even a small amount of water may not emulsify properly when added to the tank. To avoid crop injury, (4) Do not apply when crop is under stress from prolonged cold, wet weather, poor fertility, or other factors, or when crop is wet and succulent from recent rainfall. (5) Do not exceed 2.5 lbs. active ingredient (or 2.8 lbs. of this product) per acre per calendar year. (6) Postemergence applications to corn must be made before corn exceeds 12 inches in height.

Tank Mixtures for Corn

This product may be tank mixed with these herbicides for control of certain broadleaf and grass weeds in corn:

Dual® (metolachlor)
Dual + Gramoxone Extra
Dual + Roundup (glyphosate)
Dual + Princep®
Dual + Princep + Gramoxone Extra
Dual + Princep + Roundup
Bexton® or Ramrod® (propachlor)
Lasso® or Lasso EC (alachlor)
Lasso or Lasso EC + Roundup
Lasso or Lasso EC + Gramoxone Extra
Gramoxone Extra
Princep
Princep + Gramoxone Extra
Princep + Roundup
Roundup
Sutan + ®

Use tank mix directions appearing on the labels of the above herbicides when tank mixing with this product. Observe all precautions and limitations on labeling of products used in a particular tank mix.

Note: When the labels of the above herbicides refer to atrazine 80W, use equivalent rate of AAtrex Nine-O. One lb. of 80W equals 0.9 lb. of Nine-O.

Princep 80W, Princep 4L, or Princep Caliber 90®
In addition to the weeds listed under AAtrex Nine-O Applied Alone - Corn or Grain Sorghum - Preplant Surface-Applied, Preplant Incorporated, or Preemergence, this combination also controls crabgrass, fall panicum, and carpetweed.

Broadcast tank mix before planting, at planting, or after planting, but before crop and weeds emerge, at rates in Table 2. Use the 1:1 ratio for control of most weeds. Use the 1:2 ratio for expected heavy infestations of crabgrass and fall panicum. Cultivate shallowly if weeds develop.

Preplant Surface-Applied: Use on medium- and fine-textured soils with minimum-tillage or no-tillage systems only in CO, IA, IL, IN, KS, KY, MN, MO, MT, ND, NE, SD, WI, and WY. Apply the recommended rate of AAtrex and Princep shown in Table 2 up to 45 days prior to planting. Refer to the AAtrex Nine-O alone section for information if weeds should develop following early treatment. On coarse-textured soils, do not apply more than two weeks prior to planting. Refer to the AAtrex Nine-O Alone - Preplant Surface-Applied section of the corn label for additional details.

If weeds are present at time of treatment, apply in a tank mix combination with a contact herbicide (for example, Gramoxone Extra or Roundup). Observe directions for use, precautions, and restrictions on the label of the contact herbicide.

Note: To the extent possible, do not move treated soil out of the row or move untreated soil to the surface during planting, or weed control will be diminished.

Preplant Incorporated: Apply to the soil and incorporate in the spring before, during, or after final seedbed preparation. Avoid deep incorporation. For best results, apply within two weeks prior to planting.

Preemergence: Apply during or shortly after planting, but before crop and weeds emerge.

Refer to Corn sections of this label and to Princep 80W, Princep Caliber 90, or Princep 4L labels for further directions, limitations, and precautions.

Table 2: Tank Mixtures with Princep on Corn

	Broadcast Rate/A			
	1:1 Ratio *		1:2 Ratio **	
Soil Texture	This Product	Princep 80W[1]	This Product	Princep 80W[1]
Sand, loamy sand, sandy loam	1.1 lbs.	1.25 lbs.	0.73 lb.	1.67 lbs.
Loam, silt loam, silt, clay loam, sandy clay loam, silty clay loam, sandy clay, or silty clay with low organic matter	1.3 lbs.	1.5 lbs.	0.88 lb.	2 lbs.
Loam, silt loam, silt, clay loam, sandy clay loam, silty clay loam, sandy clay, or silty clay with medium to high organic matter, and clay (including dark prairie soils of the Corn Belt)	1.6 lbs.	1.8 lbs.	1.07 lbs.	2.4 lbs.

* For control of most weeds.

** For control of expected heavy infestations of crabgrass and fall panicum.

[1] When using Princep Caliber 90 or Princep 4L, use equivalent rates. One lb. of Princep 80W equals 0.9 lb. of Princep Caliber 90 or 1.6 pts. of Princep 4L.

Princep 80W, Princep 4L, or Princep Caliber 90 plus Roundup: Use as tank mixture for preemergence and postemergence control of certain broadleaf and grass weeds where corn will be planted directly into a cover crop, established sod, or in previous crop residues. Refer to Roundup label for all directions, weeds controlled, precautions, and limitations.

Princep 80W, Princep 4L, or Princep Caliber 90 plus Gramoxone Extra: Use as tank mixture with Princep and Gramoxone Extra to kill existing vegetation and for residual weed control where corn will be planted directly into a cover crop, established sod, or in previous crop residues. Add this product and Princep to water in spray tank, agitating until thoroughly mixed. Then add Gramoxone Extra and a nonionic surfactant, such as X-77®. Continue agitation during application. Broadcast 1.1-2.2 lbs. of this product plus 1.25-2.5 lbs. of Princep 80W (or 2-4 pts. of Princep 4L, or 1.1-2.2 lbs. of Princep Caliber 90) plus a suitable amount of Gramoxone Extra in 20-80 gals. of water per sprayed acre. Refer to the Gramoxone Extra label for the appropriate rates to utilize in this tank mixture. Apply before, during, or after planting, but before corn emerges. Add 0.5 pt. of a nonionic surfactant, such as X-77, per 100 gals. of spray mixture. Use the higher rate of Gramoxone Extra specified on the label if existing vegetation is 4-6 inches tall. This mixture will not control weeds taller than 6 inches.

Refer to further limitations and precautions on labels for this product, Princep, and Gramoxone Extra.

Precautions: For all applications to corn: (1) To avoid crop injury and illegal residues, do not apply more than 2.8 lbs./A of this product per year. (2) For best control of velvetleaf and cocklebur, the application rate cannot be less than 2 lbs./A active ingredient, either alone or in tank mix combinations. (3) Following harvest, plow (moldboard or disk-plow) and thoroughly till soil in fall or spring to minimize possible injury to spring-seeded rotational crops, regardless of rate used.

Note for all applications to corn: Do not graze or feed forage from treated areas for 21 days following application, or illegal residues may result.

Sorghum and Sorghum-sudan Hybrids (Grain and Forage Types)

Preplant Surface-Applied (Broadleaf and grass control): Use on medium- and fine-textured soil with minimum-tillage or no-tillage systems only in CO, IA, IL, IN, KS, KY, MN, MO, MT, ND, NE, SD, WI, and WY. Apply the recommended rate of AAtrex Nine-O shown in Table 3 up to 45 days prior to planting. If an unsatisfactory length of weed control results from adverse environmental conditions following early treatment, a follow-up application of an appropriately labeled herbicide may be used. If the follow-up treatment includes atrazine, do not exceed the labeled rate for corn indicated in Table 1. Under dry conditions, irrigation after application is recommended to move AAtrex Nine-O into the soil.

If weeds are present at time of treatment, apply in a tank mix combination with a contact herbicide (for example, Gramoxone Extra or Roundup). Observe directions for use, precautions, and restrictions on the label of the contact herbicide.

Note: To the extent possible, do not move treated soil out of the row or move untreated soil to the surface during planting, or weed control will be diminished.

Preplant Incorporated (Broadleaf and grass control): Broadcast in spring after plowing at rate shown in Table 3. Apply before, during, or after final seedbed preparation. If soil is tilled or worked after application, avoid deep incorporation. For best results, apply within two weeks prior to planting.

Preemergence (Broadleaf and grass control): Apply during or shortly after planting, but prior to weed or crop emergence at rate shown in Table 3.

Postemergence (Broadleaf and grass control): Apply at rate shown in Table 3 before weeds exceed 1.5 inches in height and before sorghum exceeds 12 inches in height.

Table 2

AAtrex® Nine-O®

Table 3: Broadleaf and Grass Weed Control in Sorghum[1],[2]

FOR ALL SOIL APPLICATIONS PRIOR TO CROP EMERGENCE

● **On Highly Erodible Soils (as defined by SCS)**

If conservation tillage is practiced, leaving at least 30% of the soil covered with plant residues at planting, apply a maximum of 2.2 lbs./A as a broadcast spray

If the soil coverage with plant residue is less than 30% at planting, a maximum of 1.8 lbs./A may be applied

● **On Soils Not Highly Erodible**

Apply 2.2 lbs./A as a broadcast spray

FOR POSTEMERGENCE APPLICATION

If no atrazine was applied prior to sorghum emergence, apply a maximum of 2.2 lbs./A broadcast. If a postemergence treatment is required following an earlier herbicide application, the total atrazine applied may not exceed 2.5 lbs. active ingredient (2.8 lbs. of this product) per acre per calendar year.

[1]Do not apply preplant surface or preplant incorporated in AL, AR, FL, GA, LA, MS, NC, NM, OK, SC, TN, or TX. Do not apply preemergence in NM, OK, or TX, except in northeast OK and the TX Gulf Coast and Blacklands areas.

[2]Where there are state/local requirements regarding atrazine use (including lower maximum rates and/or greater setbacks) which are different from the label, the more restrictive/protective requirements must be followed. Certain states may have established rate limitations within specific geographical areas. Consult your state lead pesticide control agency for additional information. It is a violation of this label to deviate from state use regulations.

In case of planting failure, sorghum or corn may be replanted. Do not make a second broadcast application, or injury may occur. If originally applied in a band and sorghum or corn is replanted in untreated row middles, this product may be applied in a band to the second planting provided the maximum application rate of 2.5 lbs. ai/A atrazine per calendar year is not exceeded.

Preemergence broadleaf weed control in furrow irrigated bedded sorghum (AZ and CA only): For preemergence control of many broadleaf weeds, broadcast 0.9-1.3 lbs./A after bed preparation, during or after planting, but before sorghum and weeds emerge and before the first furrow irrigation. Follow with several regular irrigations, making sure to thoroughly wet all soil.

Precautions for preemergence application to furrow irrigated bedded sorghum in AZ and CA: To avoid possible sorghum injury, do not use on sand or loamy sand soil or on sorghum planted in furrows. Applications to sorghum growing on alkali soils or where cuts, fills, or erosion have exposed calcareous or alkali subsoils may cause crop injury. In case of crop failure, do not replant sorghum for 8 months following application. Corn may be planted immediately.

Postemergence broadleaf weed control with emulsifiable oil or oil concentrate in water: Broadcast 1.3 lbs./A for control of many broadleaf weeds. Apply before pigweed and lambsquarters reach 6 inches in height and before all other weeds reach 4 inches in height. In CO, western KS, NM, OK, TX, and desert regions of AZ and CA, apply when sorghum is 6-12 inches in height, but before it reaches boot stage. In all other areas, apply after sorghum reaches the 3-leaf stage but before it exceeds 12 inches in height. Add 1 gal. of emulsifiable oil/A for ground application and 0.5 gal./A for aerial application, or add 1 qt. of oil concentrate for ground application. A cultivation may be necessary if all weeds are not controlled or if weeds regrow.

For the list of weeds controlled, see AAtrex Nine-O Applied Alone – Corn or Grain Sorghum – Postemergence with Emulsifiable Oil or Oil Concentrate in Water.

Precautions for applications with emulsifiable oil or oil concentrate in water. See "Precautions for applications with emulsifiable oil or oil concentrate in water" in Corn section.

Postemergence broadleaf weed control with surfactant (CO, western KS, NM, OK, TX, and desert regions of AZ and CA only): Broadcast 1.3 lbs./A plus 0.75-1.5 pts. of surfactant after sorghum reaches 6 inches in height, but before weeds exceed 1.5 inches in height. Apply only on sandy loam and finer textured soil.

Precautions: For all applications to sorghum: (1) Heavy rain immediately following application tends to cause excessive concentrations of herbicide in seed furrow, resulting in possible crop injury. Do not apply to furrow-planted sorghum until furrows are leveled (plowed in). (2) Level deep planter marks or seed furrows before application. (2) Application to sorghum growing under stress caused by minor element deficiency or to sorghum growing on highly calcareous soil may result in crop injury. (3) Following harvest, plow (moldboard or disk-plow) and thoroughly till soil in fall or spring to minimize possible injury to spring-seeded rotational crops, regardless of rate used. (4) Injury may occur if both this herbicide, preplant surface, preplant incorporated, or preemergence, and an at-planting systemic insecticide applied in-furrow are used. (5) Do not apply more than 2.5 lbs. active ingredient (2.8 lbs. of this product) per acre per calendar year. (6) For all soil applications prior to crop emergence (except for preemergence use on bedded sorghum in AZ and CA), do not apply to coarse-textured soils, i.e., sand, loamy sand, sandy loam, or to medium and fine-textured soils having less than 1% organic matter, or injury may occur. (7) For postemergence applications, do not apply to sand or loamy sand, or injury may occur.

Note: Do not graze or feed forage from treated areas for 21 days following application, or illegal residues may result.

Tank Mixtures for Grain Sorghum

Dual 8E: Use as tank mixture with Dual 8E for control of those weeds listed on the Dual 8E label, as well as on this label. Use this tank mixture only on sorghum seed treated with Concep®. Refer to the Dual 8E label for all directions, precautions, and limitations.

Winter Weed Control in Texas

For postemergence control of winter weeds only, such as henbit, seedling dock, and annual thistle on fall bedded land in the Gulf Coast and Blacklands of TX. Apply 0.9-1.1 lbs./A postemergence to the weeds in November or December to land that will be planted to corn, grain sorghum, or forage sorghum the following spring. For best results, add a suitable surfactant, such as X-77, at the rate of 0.5% of the spray volume, an emulsifiable oil at the rate of 1.0% of the spray volume, or an oil concentrate at the rate of 1 qt. per acre.

Normal weed control programs may be used in the following corn, grain sorghum or forage sorghum crop.

Note: Do not plant any crops except corn, grain sorghum, or forage sorghum the spring following this treatment, or illegal residues may result.

AAtrex Nine-O Alone – Chemical Fallow

Wheat-Sorghum-Fallow: To control annual broadleaf and grass weeds following wheat harvest and in the following sorghum crop when grown under minimum tillage, broadcast 3.3 lbs./A to wheat stubble immediately following wheat harvest. If weeds are present at application, remove them with a sweep plow or other suitable implement after application, or use an approved contact herbicide before or after the application of AAtrex Nine-O. Plant sorghum into wheat stubble the following spring with minimum soil disturbance. Use a surface planter or a planter leaving a shallow furrow. If weeds are present at planting, remove them with a sweep plow or other suitable implement before planting.

For the list of weeds controlled, see AAtrex Nine-O Applied Alone – Corn or Grain Sorghum – Preplant Surface-Applied, Preplant Incorporated, or Preemergence.

Precautions: (1) Use only on silt loam or finer textured soil or crop injury may result. (2) Wheat-sorghum-fallow cropping sequence must be followed. (3) Do not apply following sorghum harvest.

Note: To avoid illegal residues, do not graze or feed forage from treated area to livestock. To avoid illegal residues and crop injury, do not plant any crop other than those on this label within 18 months following treatment.

Wheat-Corn-Fallow (CO, KS, NE): This product controls cheatgrass (downy brome, chess), kochia, mustards, pigweed, Russian thistle, wild lettuce, wild sunflower, and volunteer wheat during period after wheat harvest. Weed control may extend into following corn crop grown under minimum tillage.

Follow directions for use, notes, and precautions in the **Wheat-Sorghum-Fallow** section above, substituting corn for references to sorghum.

Wheat-Fallow-Wheat (CO, KS, MT, ND, NE, SD, and WY): For preemergence control of cheatgrass (downy brome, chess), common lambaquarters, field pennycress, kochia, mustard, Russian thistle, wild lettuce, and suppression of volunteer wheat during fallow period of a wheat-fallow-wheat rotation, broadcast 0.5-1.1 lbs./A on all soils except those listed under *Precautions*. For control of pigweed and wild sunflower, use the higher rate. Apply to stubble ground. Treat only once during same fallow period.

Tank Mixtures for Chemical Fallow

Wheat-Sorghum-Fallow or Wheat-Corn-Fallow (KS, NE)
Gramoxone Extra: If weeds are present at application, a tank mix with Gramoxone Extra may be used. Broadcast 3.3 lbs. of AAtrex Nine-O plus a suitable amount of Gramoxone Extra in 20-60 gals. of water/A by ground equipment. Refer to the Gramoxone Extra label for the appropriate rates to utilize in this tank mixture. Add 0.5-1 pt. of a nonionic surfactant, such as X-77, per 100 gals. of spray mixture. Add AAtrex Nine-O to spray tank first and thoroughly mix with water. Then add Gramoxone Extra, followed by surfactant. Use the higher rate of Gramoxone Extra specified on the label if weeds are 4-6 inches tall. This mixture will not control weeds taller than 6 inches. Apply to stubble ground. Treat only once during same fallow period. Refer to Gramoxone Extra label for further directions, precautions, and limitations.

Wheat-Fallow-Wheat (CO, KS, ND, NE, SD, and WY)
Gramoxone Extra: If weeds are present at application, a tank mix with Gramoxone Extra may be used. Broadcast 0.5-1.1 lbs. of AAtrex Nine-O plus a suitable amount of Gramoxone Extra in 20-60 gals. of water/A by ground equipment. Add 0.5-1 pt. of a nonionic surfactant, such as X-77, per 100 gals. of spray mixture. Add AAtrex Nine-O to spray tank first and thoroughly mix with water. Then add Gramoxone Extra, followed by surfactant. Use the higher rate of Gramoxone Extra specified on the label if weeds are 4-6 inches tall. This mixture will not control weeds taller than 6 inches. Apply to stubble ground. Treat only once during same fallow period. Refer to Gramoxone Extra label for further directions, precautions, and limitations.

If weeds are present at application and this product is used alone, use either an approved contact herbicide before or after treatment, or tillage after treatment.

Use tillage to control weeds which escape during fallow period. Till before planting. For this product applied alone or in tank mixture with Gramoxone Extra, plant at least 2 inches deep and 12 months or more after application.

Table 2

AAtrex® Nine-O®

Precautions: To avoid crop injury, (1) Do not use on sand soil. (2) Do not treat eroded hillsides, caliche and rocky outcroppings, or exposed calcareous subsoil. (3) Do not treat soils of the Rosebud and Canyon Series in western NE and adjoining counties in CO and WY. (4) Do not treat soils with calcareous surface layers. (5) Avoid spray overlap.

Note: Do not graze treated areas within 6 months after application, or illegal residues may result.

Aerial application: In order to assure that spray will be controllable within the target area when used according to label directions, make applications at a maximum height of 10 ft., using low drift nozzles at a maximum pressure of 40 psi, and restrict application to periods when wind speed does not exceed 10 mph. To assure that spray will not adversely affect adjacent sensitive nontarget plants, apply AAtrex Nine-O alone by aircraft at a minimum upwind distance of 400 ft. from sensitive plants.

Roadsides

To control certain annual weeds in established perennial grasses along roadsides in CO, KS, MT, ND, NE, SD, and WY, including cheatgrass (downy brome, chess), common (annual) broomweed, little barley, medusahead, sagewort, and tumble mustard, broadcast 1.1 lbs./A in a minimum of 10 gals. of water by ground equipment in the fall before ground freezes, or after thawing in the spring, but before the established grasses green-up and before weeds emerge. Examples of desirable established grasses include big bluestem, bluegrama, bromegrass, buffalograss, crested wheatgrass, indiangrass, little bluestem, side-oats grama, switchgrass, and Western wheatgrass. Apply only once per year. Temporary discoloration or other form of injury to the desirable perennial grasses may occur following application.

Notes: To avoid illegal residues, (1) Do not cut or feed roadside grass hay. (2) Do not allow livestock to graze treated areas.

Sugarcane

For control of many broadleaf and grass weeds, including amaranths, crabgrass, fireweed, Flora's paintbrush, foxtails, junglerice and wiregrass, broadcast 2.2-4.4 lbs./A of AAtrex Nine-O at time of planting or ratooning, but before sugarcane emerges. Broadcast aerially in a minimum of 5 gals. of spray per acre, or broadcast or band by ground equipment in a minimum of 20 gals. per acre, unless indicated otherwise. One additional application may be made over the sugarcane as it emerges, and two additional applications may be made interline after emergence as directed sprays. Repeat treatments, where needed, may be applied broadcast, band, or interline as suggested with the final application being prior to close-in. Do not exceed the rate of herbicide suggested for any one crop of sugarcane.

Note: Where high rates of AAtrex Nine-O are used, apply in a minimum of 1 gal. of water for each 1 lb. of product applied per acre.

Aerial application: In order to assure that spray will be controllable within the target area when used according to label directions, make applications at a maximum height of 10 ft., using low drift nozzles at a maximum pressure of 40 psi, and restrict application to periods when wind speed does not exceed 10 mph. To assure that spray will not adversely affect adjacent sensitive nontarget plants, apply AAtrex Nine-O alone by aircraft at a minimum upwind distance of 400 ft. from sensitive plants.

Florida

For control of emerged pellitory weed: Apply 0.4-0.6 lbs./A in at least 40 gals. of water as a directed spray by ground equipment prior to close-in. Add 4 qts. of surfactant for each 100 gals. of spray. Thoroughly cover weed foliage.

For control of alexandergrass, large crabgrass, pellitory (artillery) weed, and spiny amaranth, use one of the following methods at planting or ratooning:

1. Apply 4.4 lbs./A preemergence. Follow with one or two applications, as needed, postemergence to sugarcane and weeds, at 2.2 lbs./A. Treat before weeds exceed 1.5 inches in height.

2. Apply 1-3 times, as needed, at 2.2 lbs./A postemergence to sugarcane and weeds. Treat before weeds exceed 1.5 inches in height.

Louisiana

For control of annual weeds during the summer fallow period, apply 2.2 lbs./A to weed-free beds immediately after bed formation. Follow normal weed control program after planting.

Precautions: To avoid crop injury, (1) Do not apply more than 11 lbs./A to any one crop of sugarcane. (2) If making a 2.2 lbs./A application during summer fallow period, do not exceed 8.8 lbs./A during the remainder of the growing season, or illegal residues may result.

Texas

Use AAtrex Nine-O for control of barnyardgrass, pigweed, purslane and sunflower, in plant or ratoon sugarcane.

Apply 4.4 lbs./A of AAtrex Nine-O preemergence. Follow with one or two applications, as needed, at 3.3 lbs./A postemergence to sugarcane and weeds.

For best results when weeds are emerged, add a nonionic surfactant at a concentration of 2 qts./100 gals. to the spray and apply before weeds exceed 1.5 inches in height.

Precautions: (1) Injury to sugarcane may occur when under moisture stress, when soil is of low adsorptive capacity, or when land is first cropped to sugarcane. (2) Do not apply after close-in. (3) Do not apply more than 11 lbs./A to any one crop of sugarcane, or crop injury may result.

Turfgrasses for Sod (Florida only)

St. Augustinegrass, Centipedegrass, and Zoysiagrass

Broadcast 2.2-4.4 lbs./A according to soil texture to control those weeds listed under **AAtrex Nine-O Applied Alone - Corn or Grain Sorghum - Preplant Surface-Applied, Preplant Incorporated, or Preemergence.**

Muck or peat	4.4 lbs.	Old beds: Within 2 days after lifting sod
		New beds: 3-4 days after sprigging or plugging
Sandy soil	2.2 lbs.	Old beds: Within 2 days after lifting sod
		New beds: 7-10 days after sprigging or plugging

If weeds regrow, apply an additional 2.2 lbs./A on muck or peat, or 1.1 lbs./A on sandy soil.

Precautions: To avoid crop injury, (1) Do not apply within 30 days prior to cutting or lifting. (2) Do not apply in combination with surfactants or other spray additives. (3) Use only on turfgrass reasonably free of infestations of insects, nematodes, and diseases. (4) On newly sprigged turfgrass, temporary slowing of growth may follow application.

Turfgrass for Fairways, Lawns, Sod Production [*] and Similar Areas

[*] In states other than FL. For use on turfgrass for sod in FL, see **Turfgrasses for Sod (Florida only)** section above.

Bermudagrass, Centipedegrass, St. Augustinegrass, and Zoysiagrass

Apply AAtrex Nine-O after October 1 before emergence of winter annual weeds for control of annual bluegrass, burclover, carpet burweed, chickweed, corn speedwell, henbit, hop clover, and spurweed. AAtrex Nine-O will control annual bluegrass even if it is emerged at time of treatment. For control of summer annual weeds listed in the preemergence section of the **AAtrex Nine-O Applied Alone - Corn or Grain Sorghum** section of this label, also apply AAtrex Nine-O in late winter before the weeds emerge. Apply in a minimum of 15 gals. of water per acre or 1 gal. per 1,000 sq. ft.

Where annual bluegrass is the major weed, use 1.1 lbs. of AAtrex Nine-O per acre (0.4 fl. oz. per 1,000 sq. ft.). Use 2.2 lbs./A (0.8 fl. oz. per 1,000 sq. ft.) for control of the other weeds named above. Do not exceed 1.1 lbs. per acre per treatment on newly sprigged turfgrass or on hybrid bermudagrass such as Tiflawn, Tifway, and Ormond.

For continued summer annual weed control, apply another 1.1 lbs./A at least 30 days after the previous application, but not after April 15. However, do not make more than two applications of this product per year.

Precautions: On newly sprigged turfgrass and hybrid bermudagrass, temporary slowing of growth and yellowing may occur following application. To avoid turf injury, (1) Use only on turfgrass reasonably free of infestations of insects, nematodes, and diseases. (2) Do not use on golf greens. (3) Do not use north of NC (except may be used in VA Coastal Plains) or west of the high rainfall areas of eastern OK and eastern TX. (4) Do not use on muck or alkaline soils. (5) Do not apply over the rooting area of trees or ornamentals not listed on this label. (6) Do not overseed with desirable turfgrass within 4 months before or 6 months after treatment. (7) Do not apply this product to newly seeded bermudagrass until it has overwintered and has a well-developed rhizome system. Do not exceed 2.2 lbs. product/A within 12 months of seeding bermudagrass.

Note: Do not graze or feed turf clippings to animals, or illegal residues may result.

Macadamia Nuts

For preemergence control of many broadleaf and grass weeds, including crabgrass, foxtail, wiregrass, Flora's paintbrush, spanishneedles, and fireweed, broadcast 2.2-4.4 lbs./A before harvest and before weeds emerge. Repeat as necessary. Do not spray when nuts are on ground during harvest period. Do not apply by air.

Guava

Use only on established plantings which are at least 18 months old. Apply as a directed spray at 2.2-4.4 lbs./A of AAtrex Nine-O in 20-50 gals. of spray mix preemergence or early postemergence to weeds. When applying postemergence, the use of a surfactant and greater spray volume (80-100 gals. of spray mix per acre) may enhance weed control. This product controls many annual broadleaf and grass weeds, including fireweed, purslane, scarlet pimpernel, spanishneedles, and sowthistle.

Notes: To avoid illegal residues, (1) Do not allow spray to contact foliage or fruit. (2) Do not apply more frequently than at 4 month intervals. (3) Do not apply more than 8.8 lbs. of AAtrex Nine-O per year.

Table 2

AAtrex® Nine-O®

Conifers

For control of annual broadleaf and grass weeds prior to transplanting, after transplanting or in established conifers (including Douglas fir, grand fir, noble fir, white fir, Austrian pine, bishop pine, Jeffrey pine, knobcone pine, loblolly pine, lodgepole pine (shore pine), monterey pine, ponderosa pine, Scotch pine, slash pine, blue spruce, and Sitka spruce): Broadcast 2.2-4.4 lbs. in a minimum of 5 gals. of water per acre by air or 10 gals. by ground before weeds are 1.5 inches tall. Apply to established trees between fall and early spring while trees are dormant. For new transplants, apply during or soon after transplanting. For applications prior to transplanting, allow sufficient precipitation to activate AAtrex Nine-O before transplanting. In areas where spring and summer rainfall is inadequate to activate AAtrex Nine-O, apply during fall prior to spring transplanting.

For the list of weeds controlled, see **AAtrex Nine-O Applied Alone – Corn or Grain Sorghum – Preplant Surface-Applied, Preplant Incorporated, or Preemergence.**

Quackgrass control: Broadcast 4.4 lbs. in a minimum of 5 gals. of water per acre by air or 10 gals. by ground between fall and early spring while trees are dormant and before quackgrass is more than 1.5 inches tall.

Precautions: (1) In areas west of the Rocky Mountains (except the Great Basin), grazing may begin 7 months after a fall application or 3 months after a winter or spring application. (2) To prevent illegal residues, do not graze treated areas of the Great Basin, or areas east of the Rocky Mountains. (3) Temporary injury to trees may occur following use of AAtrex Nine-O on coarse-textured soil. (4) To avoid crop injury, do not apply to seedbeds. (5) Also apply only once per year.

Aerial application: In order to assure that spray will be controllable within the target area when used according to label directions, make applications at a maximum height of 10 ft. above vegetation, using low drift nozzles at a maximum pressure of 40 psi, and restrict application to periods when wind speed does not exceed 10 mph. To assure that spray will not adversely affect adjacent sensitive nontarget plants, apply AAtrex Nine-O by aircraft at a minimum upwind distance of 400 ft. from sensitive plants.

Storage and Disposal

Pesticide Storage and Disposal

Store in a dry place. Do not contaminate water, food, or feed by storage, disposal, or cleaning of equipment. Wastes resulting from the use of this product may be disposed of on-site or at an approved waste disposal facility.

Container Disposal

Completely empty bag into application equipment. Dispose of empty bag in a sanitary landfill or by incineration, or by open burning, if allowed by state and local authorities. If burned, keep out of smoke.

For minor spills, leaks, etc., follow all precautions indicated on this label and clean up immediately. Take special care to avoid contamination of equipment and facilities during cleanup procedures and disposal of wastes. In the event of a major spill, fire or other emergency, call 1-800-888-8372 day or night.

Precautionary Statements

Hazards to Humans and Domestic Animals

CAUTION

Harmful if swallowed, inhaled, or absorbed through skin. Do not breathe dust or spray mist. Avoid contact with eyes, skin or clothing.

Users are required to wear long-sleeved shirts and long pants or equivalent, chemical resistant gloves, and boots (waterproofed). In addition, persons involved in mixing/loading operations are required to use chemical resistant rubber or neoprene gloves and a face shield or goggles.

Statement of Practical Treatment

If swallowed: Call a physician or Poison Control Center immediately. Drink 1 or 2 glasses of water and induce vomiting by touching back of throat with finger, or, if available, by administering syrup of ipecac. Do not induce vomiting or give anything by mouth to an unconscious person.

If inhaled: Remove victim to fresh air. If not breathing, give artificial respiration, preferably mouth-to-mouth. Get medical attention.

If on skin: Wash with plenty of soap and water. Get medical attention.

If in eyes: Flush eyes with plenty of water. Call a physician if irritation persists.

Note to Physician: There is no specific antidote for atrazine. If this product is ingested, induce emesis or lavage stomach. The use of an aqueous slurry of activated charcoal may be considered.

Environmental Hazards

Atrazine can travel (seep or leach) through soil and can enter ground water which may be used as drinking water. Atrazine has been found in ground water. Users are advised not to apply atrazine to sand and loamy sand soils where the water table (ground water) is close to the surface and where these soils are very permeable, i.e., well-drained. Your local agricultural agencies can provide further information on the type of soil in your area and the location of ground water.

This product may not be mixed/loaded or used within 50 feet of all wells, including abandoned wells, drainage wells, and sink holes. This product may not be mixed or loaded within 50 feet of intermittent streams and rivers, natural or impounded lakes and reservoirs. This product may not be applied aerially or by ground within 66 feet of the points where field surface water runoff enters perennial or intermittent streams and rivers or within 200 feet around natural or impounded lakes and reservoirs. If this product is applied to highly erodible land, the 66-foot buffer or setback from runoff entry points must be planted to crop, seeded with grass or other suitable crop.

This pesticide is toxic to aquatic invertebrates. Do not apply directly to water, to areas where surface water is present, or to intertidal areas below the mean high water mark. Do not apply when weather conditions favor drift from treated areas. Runoff and drift from treated areas may be hazardous to aquatic organisms in neighboring areas. Do not contaminate water when disposing of equipment wash waters.

AAtrex®, Nine-O®, Caliber 90®, Concep®, Dual® and Princep® trademarks of CIBA-GEIGY Corporation

Bexton® trademark of DowElanco

Compex® trademark of KALO Agricultural Chemicals, Inc.

Gramoxone® and Sutan + ® trademarks of ICI Americas

Lasso®, Ramrod® and Roundup® trademarks of Monsanto Company

Unite® trademark of Hopkins Agricultural Chemical Company

X-77® trademark of Chevron Chemical Company

©1992 CIBA-GEIGY Corporation

Agricultural Division
CIBA-GEIGY Corporation
Greensboro, North Carolina 27419
CGA 7L101M   052

Q & A
Von McCaskill, head of the South Carolina Department of Fertilizer and Pesticide Control and long active in the Association of American Pesticide Control Officials.

Q. What are state officials worried about?

A. States are in general very concerned about regulations coming down from EPA that will require additional training (programs that states would have to run), but no additional money to carry that out. For instance, with the new worker safety (rule), one of our major concerns is that we not have a situation develop where a worker has to be retrained in every state he works in during a season. We're working on a national verification system that gives the worker a dated card verifying that he was trained in worker safety.

Q. It seems that there are a lot of new regulations coming along that go beyond the label and require applicators to get information on their own.

A. That's true. They're moving into a lot of areas I'd say are nontraditional areas for the EPA. What you've got to do is educate the regulated community and motivate them into compliance. Because we (the states) don't have enough money to be out there enforcing these rules all the time.

The next big program coming is disposal and storage. None of that is going to be on the label. (Along with endangered species and groundwater rules) these are areas that are very important and have been overlooked for a long time. But the money (to train people and enforce the rules) is woefully inadequate. We're trying to come up with innovative ways to deal with it.

Q. Is the system for training pesticide applicators equipped to deal with these changes?

A. No, it's not even geared up to deal with worker safety. It's going to require additional efforts by the extension service and private groups. You're going to see a lot more of the burden being pushed to groups like Farm Bureau.

Q. Similar to the extensive training seminars offered by the Golf Course Superintendents Association?

A. Yes. We certainly support that concept. They're actually getting ahead of the curve. But eventually all that will be mandated by the states.

Q. Are all states moving toward more stringent regulation of applicators, such as mandatory tests for farmers or certifying commercial applicators for general use pesticides?

A. That's a difficult question to answer in generic terms. In the Southeast, for instance, many states would find it very difficult to impose a test on private applicators. In the Northeast and far West, they're moving on their own social motivations. But in the Midwest and Southeast there's just not the same kind of environmental consciousness.

Q. Are state officials still concerned about the comprehensibility of labels?

A. Yes, but it's being attacked from a different perspective. When we find individual label errors of significance they're directed to a group at EPA called the State Labeling Issues Committee. And we're working on our concerns about some termiticides — there isn't near enough efficacy data to support the use rates on the label.... In the future, with all the things going on the label referencing other information, the labels are just going to get more cluttered.

Q. How are individual applicators dealing with these changes?

A. The more sophisticated ones, such as PCOs with the pest control association, are developing supplemental materials to help them understand the new rules.

The other side of the picture are the less sophisticated ones, who are relying on county agents to make their interpretations.

Q. So are applicators in the future going to have to be better trained and educated before they get into the business?

A. There's definitely a move in that direction. Even in the farmer-grower community farmers want to be able to hire someone who has some training. A lot of this is being driven by liability concerns. There's also heightened environmental sensitivity, but these are also extremely expensive products, and they want to make sure they have someone who's going to use it right.

A lot of them are pushing very hard for (better training, certification). In South Carolina, a group of our applicators came to us asking that all technicians get verifiable training. There is an increased awareness of what can happen if someone's not trained.

Q. It seems there's a trend toward farmers just saying, "I don't want to deal with all these regulations, I'll just hire someone to do it." Is that the case?

A. Yes, where they can afford it and it's available. You see more and more of that happening across the country.

Chapter 7

STATE REGULATIONS

INTRODUCTION TO STATE SURVEY

The following chart and synopses of state certification and training programs comes from a survey of all the states and the District of Columbia conducted during the spring of 1993. It is meant to give a brief overview of the important parts of a state's pesticide regulatory program that directly affect pesticide users.

First, a couple of caveats. Note that the descriptions are not necessarily comprehensive because they depend on what and how much information the states provided in response to the survey. The survey obtained the basic information about who must be certified, how they get certified and stay certified, and how much it costs. Some states sent along extra information about their other pesticide regulations, such as notification and posting rules, and they are included when available. Also, note that these regulations are subject to change as state legislatures are apt to amend their programs at any time.

That said, this survey illustrates the wide variety of approaches different states take toward controlling the activities of people using pesticides within their borders.

The differences stem from the fact that the federal pesticides law says relatively little about exactly how certification and training is supposed to be carried out by states. It says anyone using a restricted-use pesticide must be certified, and that commercial users must prove their ability to use such pesticides safely. But states are free to create more stringent regulations if they choose.

Some states, particularly those in the Southeast, have taken their programs little farther than that, for various reasons. So if a state does not require its private applicators — basically, farmers — to take an exam before they can use restricted-use pesticides, that's usually because that state has so many farmers it would be too difficult and expensive to carry out all those exams. Also, the political situation within each state dictates a lot about its pesticide regulations.

That's been made clear in California, for instance, which has many regulatory programs that rival the federal government's. Pesticide regulation is no exception. For that reason, there is a separate chapter later in the book that describes California's program in greater detail.

Pesticide applicators and regulators agree that the trend in state regulation is toward more stringence. So some of the discretionary items in the state programs are likely to look more similar — and tougher — as time goes on.

For instance, the survey found that about two-thirds of states require private applicators to take a mandatory pass-fail test before they can be certified. And about the same proportion require commercial applicators to be certified to use any pesticide, not just those on the restricted-use list.

A number of states are starting to require applicators — especially those applying chemicals in structures — to have certain education or experience qualifications before they can be certified. In seven states, structural applicators are handled under a completely separate system from those in other categories.

Many states also use a separate licensing system to keep track of pest control businesses. Most of these require a yearly fee and registration with the state besides the training and certification of the applicators who work for them.

The length of certification ranges from 1 to 5 years, with most states using either 3 or 5 years.

There was also a wide variety of fees for certification and licensing, from nothing to $150. Most fell between $25 and $50.

The states also varied widely in the number of categories for commercial applicators. Several states had created special categories to match local conditions, such as cranberries in Massachusetts and birds on bridges in Washington.

Several states have gone ahead and created categories for private applicators as well. All states might have to do so if anticipated changes to the federal certification and training regulations take place.

For more information on how to comply with each state's regulations, contact the state official listed later in this book or call a local Cooperative Extension office.

STATE CERTIFICATION
AND
TRAINING PROGRAMS

A 50-STATE SURVEY

	Mandatory Private Exam?	Certification Lasts (Yrs.)	Fee To Certify	Hours/Credit To Renew
Alabama	No — experimental program underway	3	$10-comm.	50 points (each conf. day is 10)
Alaska	Yes	3	None	
Arizona	Yes	1	$50-comm. $30-struct.	6-comm. 3-private
Arkansas	No	5-c 3-p	$25	
California	Yes, but not pass/fail	3	$40-comm. $10-struct.	4-40
Colorado	Yes, but not pass/fail	1-3	$75-comm.	8-10
Connecticut	Yes	5	$25-comm. $5-priv.	2 work- shops

	Categories		Comm. Cert. For General Use Required?	License?	Separate Structural System?	Extra Educ./Exp. For Structural?
	Private	Commercial				
Alabama	1	12	No	Yes	No	Yes
Alaska	1	14	Yes	No	No	
Arizona	1	6	Yes	Yes	Yes	Yes
Arkansas	1	8	Yes	Yes	Yes	
California	1	6	Yes	Yes	Yes	Yes
Colorado	1	22	Yes	Yes	No	Yes
Connecticut	1	10	Yes	Yes	No	

If the information provided by a state is unclear or nonspecific on any point it is left blank.

	Mandatory Private Exam?	Certification Lasts (Yrs.)	Fee To Certify	Hours/Credit To Renew
Delaware	Yes	3	$11.50	2-12
District of Columbia	Yes	3	None	1 course
Florida	Yes	4	$75-comm. $30-private $50-strruct.	4-16
Georgia	No	5	$25-comm. $50-struct.	6-10
Hawaii	Yes	5	$10	
Idaho	No	1-5	$10-comm. $25-private	15-40
Illinois	Yes	3	$50-comm. $10-private $40-struct.	16-struct.
Indiana	Yes	5	$30-license	

	Categories		Comm. Cert. For General Use Required?	License?	Separate Structural System?	Extra Educ./Exp. For Structural?
	Private	Commercial				
Delaware	1	12	Yes	Yes	No	
District of Columbia	1	8	Yes	Yes	No	Yes
Florida	1	10	No-comm. Yes-struct.	Yes	Yes	Yes
Georgia	1	14	Yes	Yes	Yes	Yes
Hawaii	1	11	No	No	No	
Idaho	1	22	Yes	Yes	No	
Illinois	1	6	Yes	Yes	Yes	Yes
Indiana	1	11	Yes	Yes	No	Yes

If the information provided by a state is unclear or nonspecific on any point it is left blank.

	Mandatory Private Exam?	Certification Lasts (Yrs.)	Fee To Certify	Hours/Credit To Renew
Iowa	Yes	3	$75-comm. $15-private	6 per year
Kansas	Yes Open book	3-comm. 5-private	$35-comm. $10-private	
Kentucky	No	5	$25-comm.	2 courses
Louisiana	Yes	3	$15-comm. $10-private	training meeting
Maine	Yes	5-comm. 3-private	$10-comm. $6-private	6-15
Maryland	Yes	3	$50-comm. $7-private	1 session/ year
Massachusetts	Yes	5	$50-comm. $25-private	

	Categories Private	Commercial	Comm. Cert. For General Use Required?	License?	Separate Structural System?	Extra Educ./Exp. For Structural?
Iowa	1	10	Yes	Yes	No	
Kansas	1	10	Yes	Yes	No	
Kentucky	1	12	Yes	Yes	Yes	Yes
Louisiana	1	10	No	Yes	No	Yes
Maine	1	26	Yes	Yes	No	Yes
Maryland	1	12	Yes	Yes	No	Yes
Massachusetts	9	20	Yes	Yes	No	No

If the information provided by a state is unclear or nonspecific on any point it is left blank.

	Mandatory Private Exam?	Certification Lasts (Yrs.)	Fee To Certify	Hours/Credit To Renew
Michigan	Yes	3	$50-comm. $10-private	
Minnesota	Yes — open book	3	$50-comm. $30-private	
Mississippi	No	3-comm. 5-private	None	1 course or exam
Missouri	No	3-comm. 5-private	$50-comm. $25-noncomm.	None
Montana	Yes	4-comm. 5-private	$75-comm. $50 private	12-comm. 6-private
Nebraska	No	3	0	0
Nevada	Yes	4	$5	

	Categories		Comm. Cert. For General Use Required?	License?	Separate Structural System?	Extra Educ./Exp. For Structural?
	Private	Commercial				
Michigan	6	27	Yes	Yes	No	Yes
Minnesota	1	16	Yes	Yes	No	Yes
Mississippi	1	13	No	Yes	Yes	Yes
Missouri	1	14	No	Yes	No	Yes
Montana	1	16	Yes	Yes	No	No
Nebraska	1	13	No	No	No	No
Nevada	3	12	No	Yes	No	Yes

If the information provided by a state is unclear or nonspecific on any point it is left blank.

	Mandatory Private Exam?	Certification Lasts (Yrs.)	Fee To Certify	Hours/Credit To Renew
New Hampshire	Yes	5	$5	4-8 or exam
New Jersey	Yes-2	5	$75-comm.	8-24
New Mexico	Yes	5	$50-comm. $5-private	20
New York	Yes	5-comm. 3-private	$15	3 courses comm.
North Carolina	No	3-5	$30-comm. $6-private	3-10
North Dakota	Yes open book	3-comm. 5-private	$30-comm. $10-private	
Ohio	Yes	3	$20-100	1

	Categories		Comm. Cert. For General Use Required?	License?	Separate Structural System?	Extra Educ./Exp. For Structural?
	Private	Commercial				
New Hampshire	1	11	Yes	Yes	No	Yes
New Jersey	1	11	Yes	Yes	No	No
New Mexico	1	11	Yes	Yes	No	Yes
New York	6	11	Yes	Yes	No	
North Carolina	1	13	Yes	Yes	Yes	Yes
North Dakota	2	12	No	No	No	No
Ohio	13	30	Yes	Yes	No	No

If the information provided by a state is unclear or nonspecific on any point it is left blank.

	Mandatory Private Exam?	Certification Lasts (Yrs.)	Fee To Certify	Hours/Credit To Renew
Oklahoma	Yes — open book	5	$25-comm.	5-20 or test
Oregon	Yes	5	$25-private	12-40
Pennsylvania	Yes	3	$50-comm.	
Rhode Island	Yes	5	$45-comm. $20-private	1
South Carolina	No	5	$25-comm. $1-private	10-comm. 5-private
South Dakota	No	2	$25	1 course
Tennessee	No	5	$20	12-28 points
Texas	Yes	3-comm. 5-private	$150-comm. $50-private	19

	Categories		Comm. Cert. For General Use Required?	License?	Separate Structural System?	Extra Educ./Exp. For Structural?
	Private	Commercial				
Oklahoma	1	20	Yes	Yes	No	No
Oregon	1	20	Yes	Yes	No	No
Pennsylvania	1	25	Yes	Yes	No	No
Rhode Island	1	10	Yes	Yes	No	No
South Carolina	1	12	No	Yes	No	No
South Dakota	1	16	Yes	Yes	No	No
Tennessee	1	13	No	Yes	No	Yes
Texas	18	23	No	Yes	Yes	Yes

If the information provided by a state is unclear or nonspecific on any point it is left blank.

	Mandatory Private Exam?	Certification Lasts (Yrs.)	Fee To Certify	Hours/Credit To Renew
Utah	No	3	$10	Exam
Vermont	Yes	5	$20-comm.	
Virginia	Yes	2	$35-comm. $15-reg tech	
Washington	Yes	5	$142-c license $23-p license	40-comm. 20-private
West Virginia	Yes	3	$10/yr-comm. $10/3 yrs-priv.	10
Wisconsin	Yes open book	5	$30-comm. $20-private	
Wyoming	No	3-comm. 5-private	$10-comm.	24

	Categories Private	Commercial	Comm. Cert. For General Use Required?	License?	Separate Structural System?	Extra Educ./Exp. For Structural?
Utah	1	12	Yes	Yes	No	No
Vermont	6	11	Yes	Yes	No	No
Virginia	7	12	Yes	Yes	No	Yes
Washington	1	30	Yes	Yes	No	No
West Virginia	2	13		Yes	No	Yes
Wisconsin	1		Yes	Yes	No	No
Wyoming	1	10	No	Yes	No	No

If the information provided by a state is unclear or nonspecific on any point it is left blank.

ALABAMA

Alabama's Department of Agriculture and Industries requires commercial applicators to take an exam to be certified. Private applicators must undertake training.

The state is trying out an experimental program in some counties for giving private applicators such as farmers a test before they can use restricted-use pesticides.

It costs $10 per commercial exam. There is no fee for private applicators at this time.

Both commercial and private applicators must be recertified every three years.

Alabama does not require certification for general-use pesticides.

Commercial applicators are broken down into categories.

Permits are given in these categories: custom pesticide applicator, custodial pesticide applicator, governmental pesticide applicator, resident pesticide applicator and pesticide consultant.

They obtain their permits in one or more of the following categories: agricultural pest control, either plant or animal; forest pest control; ornamental and turf pest control; seed treatment; aquatic pest control; right-of-way pest control; industrial, institutional, structural and health-related pest control; fumigation pest control; public health pest control; regulatory pest control; demonstration and research pest control; or aerial applicator.

The exams include questions about label and labeling comprehension, safety such as toxicity and the need for protective clothing and equipment, environment, pests, pesticide products themselves, equipment and application techniques.

Private applicators renew their permits every three years the same way they originally got them.

Commercial applicators are renewed every three years using a point system for accrual of continuing education during that time.

There is also a set of rules for professional applicators, who must also pass an exam to do work on structures.

Permits for carrying out this work are given to people passing an examination in one of the following categories: 1) control of wood-destroying organisms; 2) industrial, institutional and household pest control; 3) fumigation pest control, such as structural; 4) ornamental and turf pest control; 5) landscape, horticultural and floricultural work, a category itself broken down into two parts: a) landscape horticulturist and b) landscape planter; and 6) tree surgery.

Beyond getting a permit to use pesticides as a professional in those categories, the Alabama rules allow for people to be certified for structural pest control as a Certified Operator or Branch Supervisor. For those certifications, applicants must have passed the appropriate examination, have at least one year of working experience in the field or a college degree including entomology.

There is also a certification as supervisor for ornamental and turf pest control, landscape horticultural and floricultural work and tree surgery. Those applicants must submit a written statement outlining their training and experience in professional work.

Alabama also has a set of regulations covering application of pesticides by aircraft.

SAFER PESTICIDES

The Environmental Protection Agency is proposing changes to help bring safer pesticides to the market. Under the proposal, EPA would streamline the registration process for candidate pesticides, thereby lowering barriers to market entry. Incentives would include waiving certain fees, reducing or deferring data requirements, allowing alternative test methods and permitting use of surrogate data.

EPA would publish a list of higher-risk pesticides and their use combinations to focus public attention on them.

From September 1992 Golf Course Superintendents Association newsletter

ALASKA

In Alaska, both private and commercial applicators must be trained and pass an exam.

Applicators must be recertified every three years and they pay no fee for certification. However, there are variable charges for coursework.

Certification is provided under the following categories:

—regulatory pest control

—demonstration and research pest control and pesticide consultants

—agricultural pest control

—ornamental and turf pest control

—seed treatment

—aquatic pest control

—industrial, institutional, structural and health related

—public health pest control

—right-of-way pest control

—mosquito and biting fly pest control

—aerial

—forest

—wood preservatives and inhibiting paint

—restricted-use dealer certification

Anyone wanting to be certified must be at least 18 years old and demonstrate knowledge of the area through an oral or written exam OR a training session or correspondence course approved by the department.

In addition, commercial pesticide applicators must keep accurate records of all uses of restricted-use pesticides for at least two years after the application. The records must include the EPA registration number of the pesticide, the place where it was purchased, the amount of pesticide bought and the use made of it, when and why. Also to be recorded are the name of the applicator, the place of application and additional information about fumigations.

Further requirements state that commercial applicators can only apply pesticides in the category they're certified for; must maintain a uniform mixture in the application equipment and thoroughly clean equipment afterward.

ARIZONA

AGRICULTURAL

Arizona's approach to certification is to have both private and commercial applicators prove themselves yearly, but not necessarily with a test. Instead, applicators can take a certain number of hours of continuing education classes to renew their certification.

To be certified initially, both farmers and professionals must take and pass by 75 percent a core exam. Commercial applicators must also take an exam in the specific area in which they are interested in working. There are six of them: ag pest control, forest pest control, seed treatment, aquatic pest control, M-44 regulatory pest control and rodent regulatory pest control.

They are recertified each year by taking continuing education classes, three hours a year for private applicators and six for commercial.

The courses are put on by private commercial entities, although some are offered by the University of Arizona Council for Environmental Studies and some community colleges. Fees vary for the courses.

There is a $50 fee for certification.

The state also has a separate system of licenses it issues for growers, pesticide sellers, agricultural aircraft pilots, custom applicators, custom applicator equipment and agricultural pest control advisors.

To get licenses for those jobs, a separate examination must be passed, a fee of between $25 and $100 paid, and other specific requirements met for that particular category.

Arizona also has separate regulations for pest control advisors and regulated growers that requires them to fill out a form called Form 1080 that details what they want a custom applicator to do. Once the work is done, the applicator must certify he did it and send the Form 1080 along to state pesticide officials.

Also, applicators — including private ones — must keep track of when they use restricted pesticides on paper and hold onto those records for two years.

STRUCTURAL

The Arizona Structural Pest Control Commission certifies commercial pest control applicators in the following categories: general pest control, wood-destroying organisms, weed control, fumigation, turf and ornamental wood preservatives, aquatic, public health and right-of-way.

Arizona state law requires that each structural commercial pest control pesticide applicator become certified for both general and restricted use pesticides.

There is a $30 fee for the test. The core exam — which covers knowledge of the pesticide label, safety, environment, equipment, application techniques, laws and regulations — must be passed by 70 percent or better.

A separate business license must be obtained by anyone wanting to run their own structural pest control business.

To renew certification each year, the applicator must pay a fee and every two years show proof of having taken at least 12 hours of continuing education classes.

ARKANSAS

Arkansas gives a test to commercial applicators but not to private ones.

However, private applicators are trained on the use of pesticides by a county extension agent.

Private applicators must renew the certification every five years, while commercial ones must do so every three years.

It costs $25 to take the test.

Arkansas issues licenses for commercial, noncommercial and private applicators. Commercial applicators must show proof of financial responsibility.

The state also has special restrictions on 2,4-D, 2,4-DB and several other pesticides.

PESTICIDES AND KIDS

"In summary, better data on dietary exposure to pesticide residues should be combined with improved information on the potentially harmful effects of pesticides on infants and children. Risk assessment methods that enhance the ability to estimate the magnitude of these effects should be developed, along with appropriate toxicological tests for perinatal and childhood toxicity. The committee's recommendations support the need to improve methods for estimating exposure and for setting tolerances to safeguard the health of infants and children."

Conclusions of the National Academy of Sciences 1993 report on Pesticides in the Diets of Infants and Children.

CALIFORNIA

This is one of several states that has separate agencies to handle agriculturally related pesticide applicators and those who do structural pest control.

AGRICULTURAL

The California Environmental Protection Agency's Department of Pesticide Regulation handles licensing and certification of agricultural applicators.

The state gives an exam to commercial applicators to test their knowledge of laws and regulations regarding pesticide use, as well as safety precautions. They must also take a separate test in the specific pest control category. Those categories are qualified applicator licensee, qualified applicator certificate holder, journeyman pest control, aircraft pilot, structural pest control operator, and field representative.

Commercial applicators must pass the above exams with a score of at least 70 percent to get a license or certificate.

There is also an exam for private applicators, but it is less demanding and is not pass/fail. Instead, any private applicator wanting to use a restricted-use pesticide must take the exam and then discuss the results with the county agricultural commissioner's staff, who administer the test. A permit to possess and use restricted-use pesticides is issued once the staff is satisfied that the applicator understands the responses and is informed of regulatory, environmental and health issues.

There is no charge for the exam given to private applicators.

Commercial applicators, however, pay a $40 initial application fee, which covers the basic exam and one category. Fee for the pilot exam is $30, for the qualified applicator certificate holder, $25, and $50 for advisors and consultants. Further category exams or reexaminations cost $15 each.

Except for advisors, there is no coursework required to be certified the first time, and the state doesn't offer any. However, to be recertified in ensuing years there is a schedule of continuing education required. Commercial applicators must be recertified every two years.

Depending on which type of license or certificate they have, applicators need between four and 40 hours of approved continuing education coursework. For instance, the qualified applicator certificate and license needs two hours of laws and regulations and another two in another area, such as labeling, identification of environmentally sensitive areas, or integrated pest management.

On the other hand, an agricultural pest control adviser must keep up more on coursework, taking four hours on laws and regulations and another 36 in other areas, such as those mentioned.

Beyond certifying applicators, California requires further training through separate regulations governing employers. They must provide worker safety training for employees handling any pesticide. They must also meet hazardous communications requirements for employees. If a person's business is "pest control," they must be licensed as an agricultural pest control business or as a maintenance gardener pest control business by the state before applying any pesticide. The business must have a qualified applicator licensee with the appropriate category in order to operate.

In keeping with its image as the leading pesticide regulatory agency in the nation (including EPA), California's rules go on to require 100 percent use reporting. That means every time a pesticide is used on an agricultural commodity it must be reported to the county agricultural commissioner. Reports are also submitted for non-production use, such as those in cemeteries, parks, ditchbanks, right-of-ways and seed treatment.

Also, in California adjuvants — efficiency enhancers such as emulsifiers, spreaders and wetting agents — are considered pesticides, even though the federal government doesn't consider them thus.

STRUCTURAL

California's Department of Consumer Affairs has a Structural Pest Control Board that oversees certification and licensure of people doing structural pest control.

That board gives out applicator's certificates, operator's licenses and field representative's licenses once a written examination is passed with at least 70 percent of the answers correct.

These requirements apply not just to people using restricted pesticides in structures — they apply to use of all types of pesticides.

Applicator certificates expire every three years and are nonrenewable. There is a $10 fee for the examination and certificate.

To be licensed as a field representative, it costs $10 for the license and $10 for the exam. It can be renewed every three years.

The operator's license costs $150 and the exam $25. It can also be renewed every three years.

The structural board breaks down operators and field representatives into four categories: fumigation, general pest, termite, and wood roof cleaning and treatment.

Field representatives must prove they have training and experience in various pest related areas. People wanting to get operator's licenses must go farther, and both prove they have between two and four years experience, depending on the category, and that they've taken board-approved courses in areas such as rules and regulations, contract law, business practices and pesticides.

The state's Structural Pest Control Act includes a variety of other requirements for applicators about how they do business.

COLORADO

Colorado's Department of Agriculture, Division of Plant Industry carries out regulation of commercial applicators, whether of agricultural or structural products. The EPA, as it does in the cases of just two states, handles regulation of private applicators in Colorado.

Unlike the states themselves, EPA cannot be more stringent than what the federal pesticides law sets out for certification. That means that for private applicators in Colorado, there is no mandatory test. The EPA sends out a study course with a multiple choice, true-false exam that is mailed back to EPA. The agency sends the test back to the applicator if he hasn't gotten 80 percent of the answers right, but must certify him the second time whether he's hit 80 percent or not. There is no fee for taking the test and there are about 13,000 private applicators certified under it.

The private certification lasts four years, at which point the mail-home test is taken again.

The state has several different types of applicator licenses and registrations.

Commercial applicators: Any business applying any pesticide commercially must be licensed as a commercial applicator.

Limited Commercial or Public Applicator: Any limited commercial or public applicator applying restricted-use pesticides must be registered or may choose to be registered.

Qualified supervisor: Any individual in the employ of a commercial, limited commercial or public applicator who without supervision, evaluates pest problems or recommends pest controls using pesticides or devices or mixes, loads or applies any pesticide or sells any application services or operates devices or supervises others must licensed as a qualified supervisor.

Certified Operator: Any individual who uses restricted-use pesticides without the on-site supervision of a qualified supervisor must be licensed as a certified operator.

Each of those four categories carries with it different requirements and fees.

Exams are required for qualified supervisors and certified operators.

The state has 22 different sub-categories for agricultural, ornamental and structural pest control.

There are also specific requirements for each license, including education and field experience.

Commercial applicator licenses expire each year, as do limited commercial and public applicator registrations.

Qualified supervisor and certified operator licenses expire every three years.

Colorado has an extensive set of training requirements for renewals. In general, to renew without an exam the qualified supervisor or certified operator must obtain the following continuing education credits: 2 in the applicable state, federal and local laws and regulations; 1 in pesticides and their families; 1 in applicator safety; 1 in public safety; 1 in environmental protection; 1 in pesticide use; and 1 in each licensed category in pest management except for residential/commercial pest control, which requires 2 credits in pest management.

OTHER REGULATIONS

In addition, Colorado has a number of laws that go beyond what the federal government requires.

Some of them cover recordkeeping, equipment identification and storage.

Colorado also has a registry of people sensitive to pesticides. Applicators must take reasonable actions to give notice of the date and approximate time of any turf or ornamental pesticide application beforehand to any person residing on property nearby and whose name is on the registry. The applicator has to try at least twice to reach that person, no later than 24 hours before the application.

Colorado also has rules about notification of the public of pesticide applications. At least one sign must be put up for a lawn or yard to have a turf or ornamental application. On recreational or common property such as golf courses and playgrounds, the signs must be posted immediately adjacent to areas where the pesticides were applied. The same goes for any aquatic application.

---

30 YEARS AFTER RACHEL CARSON

Pesticides are chemicals or biological substances used all over the world to destroy or control unwanted plants, insects, fungi, rodents, bacteria and other pests. Pesticide use has doubled since the publication of "Silent Spring," increasing from some 500 million pounds per year in 1964 to over 1 billion pounds in 1989.

Approximately 25,000 pesticide products containing some 750 active ingredients are registered on the market today; 19,000 of these products need to be reregistered. While the agricultural sector is by far the major user of pesticides, accounting for 75 percent of the volume used, pesticides are also used in many other places, such as hospitals, restaurants, public parks and the home.

From testimony before Congress by the General Accounting Office, July 1992.

CONNECTICUT

Along with its rules on certification of individual applicators, Connecticut requires that all pesticide application businesses register with the Department of Environmental Protection. That includes such professionals as exterminators, lawn services, arborists, etc. There is a $60 registration fee if the business employs more than one certified applicator.

Farmers who wish to use restricted pesticides to are required to take a written test so they can be certified as private applicators.

The examination is based on two publications: a training manual and a study manual on one of ten different areas, such as vegetable, potatoes or livestock.

The test fee for private applicators is $5.

In its separate system for commercial applicators, Connecticut splits up its certification into supervisory and operational categories.

The state offers examinations in ten different areas for each type of certification. Each costs $25.

Supervisors passing the written test must also take an oral exam.

Certification lasts five years. To be recertified, the applicator must have gone to two workshops during that period.

OTHER RULES

Connecticut's pesticide laws go beyond those of the federal government in several different ways.

For one, the state has a more extensive list of restricted-use and banned pesticides.

Also, Connecticut has a number of rules on notifying the public of certain types of outdoor applications.

Commercial pesticide applicators making non-agricultural outdoor applications within 100 yards of any property line must post signs notifying the public no farther apart than every 150 feet.

The signs must be a certain size, material and color.

There are similar rules for pesticide applications to golf courses and lakes and ponds.

Connecticut also maintains a pesticide application notification registry, so that anyone wanting to know when property abutting theirs is being sprayed, they can be informed.

DELAWARE

In Delaware, anyone buying or using a restricted-use pesticide must be certified and licensed by the Delaware Department of Agriculture, unless they work under the direct supervision of a licensed applicator.

Pest Control Operators must be certified (or work under supervision) whether they apply restricted-use or general-use pesticides.

Private applicators are certified by passing a core exam. Commercial applicators must pass the core exam as well as a category exam.

The University of Delaware publishes a core manual and category manuals to study from. There is also a quarterly training review run by Cooperative Extension, which covers sessions on a number of topics, from labels to disposal and symptoms of poisoning.

Certification for private applicators lasts for three years. After that, applicators can either retake the exam or tally up the required number of recertification hours in courses.

Private applicators need three recertification hours every three years.

Commercial applicators need two hours for seed treatment; four hours for ag animal, forest, aquatic, right-of-way, fumigation, wood preservative, and regulatory categories; eight hours for ag plant, ornamental and turf, public health and demonstration and research certifications; and 12 hours for structural.

Recertification hours for private applicators can be had from quarterly training reviews, some commodity group meetings, and professional seminars. Commercial applicators can be recertified through attendance at some commodity group meetings and professional seminars.

The department will also accept requests from applicators for credit from talks they attended that they think should count toward recertification.

Commercial applicators must renew their applicator certification each year. They have three years to obtain the first group of recertification credits.

DISTRICT OF COLUMBIA

Interestingly, the District of Columbia doesn't have to distinguish between commercial and private applicators because it doesn't have any private applicators. There's no farming within the urban district's borders.

Still, its regulations technically include private applicator certification rules.

Anyone wanting to be certified as a commercial applicator must pass a written core exam as well as one in a particular specialty. The applicator must also have a year's experience.

There's no fee to take the test, although it costs $10 to take it again.

To be recertified after three years, the applicator must take an approved course.

The district's rules apply to those using either restricted-use or general-use pesticides.

All employees applying pesticides commercially must be registered with the DC government.

Each business providing services for the eradication of pests in the district must obtain a license.

PESTICIDE USE

There are approximately 20,000 currently registered pesticide products, formulated from about 700 active ingredients (these data reflect sizable cancellations by registrants in recent years). EPA estimates total user expenditures in the United States for pesticides at $7.38 billion for 1.13 billion pounds of active ingredients in 1989.

Agriculture is by far the major user of pesticides, accounting for 75 percent of volume. The remainder is divided among industrial, commercial and governmental users (18 percent) and home and garden users (6 percent).

Congressional Research Service report, August 1992

FLORIDA

STRUCTURAL

Florida's Department of Agriculture and Consumer Services, Bureau of Entomology and Pest Control, regulates structural applicators.

It has various types of certification:

Commercial pest control operators: All pesticide applications provided by pest control businesses must be supervised by a certified pest control operator certified in the category of pest control being performed. The categories are: general household pest control, lawn and ornamental pest control, termite and wood-destroying organisms pest control and fumigation.

Certification fee is $150 per category.

Limited certification for governmental pesticide applicators and private applicators: This is for persons who apply any pesticides on government or private properties where the public may be exposed to the applications. Private applicator in this context does not refer to farmers. Categories are limited certification-structural and limited certification-lawn and ornamental.

Certification fee is $50 per category.

Limited certification for commercial landscape maintenance personnel: This is for people who are primarily in the lawn maintenance business but also wish to be certified for some pesticide application on the lawns that they maintain.

Fee is $50.

Public health pest control: Mosquito and biting fly control applicators are required to be certified in this category.

There is no fee for this category.

Commercial pest control operator and limited certification commercial landscape maintenance personnel require annual recertification. Public health pest control and limited certification for government pesticide applicators and private applicators both require recertification only once every four years.

All these structural certifications cover both general and restricted-use pesticides.

AGRICULTURAL

Florida's Department of Agriculture and Consumer Services' Bureau of Pesticides certifies and licenses applicators who purchase and/or use restricted-use pesticides in predominantly agricultural areas.

Private applicators take a core test and pay a $30 license fee for a four-year period. To be recertified they must take four continuing education credits.

If a private applicator applies pesticides aerially he must take an additional aerial exam and get 16 continuing education credits to recertify.

Public and commercial applicants take the core test and one or more primary category exams depending on where they will use restricted-use pesticides. In addition, these applicators must also take the exam for any secondary category that may apply to their use. Public and commercial applicators must test in every category that will apply to their use of restricted-use pesticides.

Public applicators pay $30 for the license and commercial ones pay $75.

To be recertified and licensed a licensed applicator must accumulate the required number of continuing education requirements or go through the examination process again.

Florida also has special rules on specific pesticides such as aldicarb, methyl bromide and alachlor.

The Bureau of Pesticides also licenses dealers of restricted-use pesticides. At this time there is not a certification or exam process associated with this license.

INDIVIDUAL RESPONSIBILITY

"There are a lot of things we as individuals have to change. Something that I wish more people would do is rather than demand that someone else solve an environmental problem, we give equal time to what we, as individuals can do as well. It's fine to sit there and say that some big company should stop polluting or that 'the government' has to do thus and so. But when are we gong to stop, in our individual daily lives, being the biggest polluters on Earth?''

John A. Moore, former EPA assistant administrator for pesticides and toxic substances

GEORGIA

AGRICULTURAL

Georgia has 14 categories for commercial applicators to receive certification to apply restricted-use pesticides: agricultural plants, agricultural animals, forest, ornamental and turf, seed treatment, aquatic, right-of-way, public health, regulatory, demonstration and research, aerial along with a category for industrial, institutional, structural and health-related applicators. There are also agricultural commodity fumigators and antifoulant paint users.

There is a single category for certification of private applicators.

Commercial applicators must pass a test to be certified. Private applicators must "demonstrate a practical knowledge of pest problems and pest control practices" with competence being verified by a written or oral test administered by the agriculture commissioner OR by attendance at and evidence of participation in a training program approved by the commissioner.

Commercial applicator licenses are good for five years. After that, recertification can be done by taking the test again or by completing between six and 10 hours of training, depending on which category he is licensed in. Recertification costs $25.

For private applicators, certification also lasts for five years.

After that, the applicator can either take three hours of training or pass another exam or its equivalent. There is no fee for recertification.

Georgia also requires its licensed pesticide contractors to keep records of all applications of restricted-use pesticides.

The law sets out that uncertified people may apply restricted-use pesticides under the direct supervision of a certified applicator. However, that doesn't mean the certified applicator has to be there at the time, so long as the people he is supervising have specific instructions. It acknowledges, though, that some labels may specify that only certified applicators can use them.

STRUCTURAL

Georgia's Department of Agriculture has a Structural Pest Control Commission that handles certification of people applying pesticides in buildings.

There are a set of fees for a company license ($50), a research fee for each company license that goes to the University of Georgia's entomology department ($90), a fee for operator certification ($50) and for each employee registration ($10).

To be certified, an applicator must have two years of actual service experience in the structural pest control category in which certification is sought; or a degree from a recognized college or university with advanced training or a major in entomology, sanitary or public health engineering and one year of actual service experience.

Once deemed qualified, the applicant must then take both oral and written exams.

Under a new regulation taking effect July 1, 1993, each pest control operator must have 10 hours of classroom training in six areas: state and federal laws and regulations, how to read and interpret a pesticide label, handling of emergencies and spills, proper methods of handling pesticides, safety and health issues and potential adverse effects on the environment.

The new rules also require 70 hours of on-the-job experience under the constant personal supervision of a certified operator or registered employee.

To be recertified, pest control operators in Georgia must either pass a written exam or accumulate a total of 10 creditable hours of training.

Georgia's pest control commission also has a detailed set of regulations setting out how inspections are to be done and standards for certain fumigations and other applications.

EPA CARCINOGEN GUIDELINES

Category A, Human Carcinogen: "Sufficient" evidence of carcinogenicity in humans from epidemiology studies to support a causal association between exposure to the agents and cancer.

Category B, Probable Human Carcinogen: B1 - "Limited" evidence of carcinogenicity in human studies and "sufficient" evidence of carcinogenicity in animals.

B2 - "Inadequate" or "no" evidence of carcinogenicity in humans and "sufficient" evidence of carcinogenicity in animals.

Category C, Possible Human Carcinogen: "Limited" evidence of carcinogenicity in animals and the "absence of human data."

Category D, Not Classifiable as to Human Carcinogenicity: "Inadequate" or "no" human and animal evidence of carcinogenicity.

Category E, Evidence of Non-Carcinogenicity for Humans: "No" evidence of carcinogenicity in two different animal species or in one animal and one human study.

HAWAII

Hawaii's Department of Agriculture has 11 categories of commercial applicators for the purposes of certification: agricultural, forest, ornamental and turf, seed treatment, aquatic, right-of-way, public health, regulatory, demonstration and research, aerial and industrial, institutional, and structural.

Commercial applicators must take both a core exam and a test in the appropriate category.

Private applicators must also take an exam, though the range of their knowledge does not have be nearly as broad as that of commercial applicators.

Certifications are good for five years. There is a test fee of $10.

Hawaii also has some brief regulations regarding posting of signs. It simply states that the department head may require the posting of signs in an area treated with a highly toxic pesticide.

HIGHTOWER SPEAKS

"There's not near enough (training of pesticide applicators). It's not a matter of labels, it's a matter of having training programs and a right-to-know program for everybody: applicators, farmers, consumers, workers. And as much as possible putting power in the hands of the people themselves for direct action.

"We (in Texas) put together a very comprehensive right-to-knowprogram and imposed new regulations to protect workers onre-entry into fields, notification of neighbors and farmers and workers that spraying was about to happen, which just created an uproar with the Farm Bureau and the chemical lobby. But we got it done. We set up our own training program. It's the only way it's going to work."

Jim Hightower, populist commentator and Texas Agriculture Commissioner from 1983 to 1991.

IDAHO

Idaho's Department of Agriculture has several different categories for licensing:

1. Dealer. Someone who represents a firm that sells restricted-use pesticides. Must pass by at least 70 percent score a written exam for pest control consultants and pay $50 for the license, which must be renewed annually.

2. Pest control consultant. Makes recommendations or supplies technical advice concerning the use of any pesticide for agricultural purposes. However, this license does not allow the holder to apply pesticides. To get one, the applicant must pass the exam by 70 percent and pay $50.

3. Private applicator. Anyone growing agricultural commodities who uses or supervises the use of restricted-use pesticides on his own land or on land rented by him or his employee. Must attend a University of Idaho Cooperative Extension Service private applicator training session OR pass a private applicator exam; and pay $25 for a license, good for five years.

4. Commercial Applicator. Someone who owns or operates a business engaged in applying pesticides to the land or property of another. Must be licensed regardless of whether he applies general-use or restricted-use pesticides. Must pass by 70 percent exams in the appropriate category or categories, provide written proof of financial responsibility, submit $50 for a license that must be renewed annually and submit a $25 fee for each piece of application equipment that must be licensed.

5. Commercial operator. Works for a commercial applicator and can only be licensed in the categories that the commercial applicator is licensed in. Must work for a licensed commercial applicator, pass an exam by 60 percent, and pay $40 for a yearly license.

6. Mixer-loader. Works for a commercial applicator. Must work for one who is licensed; also must either receive approved training or pass an exam with at least a 60 percent score.

7. Limited applicator. For those who use or supervise the use of restricted-use pesticides, but do not offer services to the public for hire and are not directly involved in commodity production. Examples are government agency employees or employees of storage or production facilities. Must pass an exam by 70 percent and pay $50 for a yearly license.

Idaho gives 22 different types of exams. There is an examination fee of $10 per category.

Idaho has reciprocity agreements with Montana, Wyoming, Utah, Oregon and Washington, which exempts them from having to take the exam in Idaho, with certain exceptions.

To be recertified, private applicators must have 15 approved training credit hours or pass a recertification exam.

All other categories must have 40 approved credit hours or pass the appropriate recertification exams. These are different from and tend to be more difficult than the initial exams.

Government agency employees are exempt from paying exam and licensing fees, but still must meet certification requirements.

Under a separate law, individuals who apply fertilizers or pesticides through irrigation water (chemigation) must certify to do so through examination or training and obtain a yearly license.

INDUSTRY CONTACTS

National Agricultural Chemicals Association, 1155 15th St. NW, Ste. 900, Washington DC 20005; telephone (202) 296-1585.

Chemical Specialties Manufacturers Association, 1913 I St. NW, Washington DC 20006; telephone (202) 872-8110.

National Agricultural Aviation Association, 1005 E St. SE, Washington DC 20003; telephone (202) 546-5722.

National Arborist Association, The Meeting Place Mall, Route 101, P.O. Box 1094, Amherst, NH 03031-1094; telephone (603) 673-3311.

National Pest Control Association Inc., 8100 Oak St., Dunn Loring, VA 22027; telephone (703) 573-8330.

Professional Lawn Care Association of America, 1225 Johnson Ferry Rd, Ste. B-220, Marietta, GA 30068; telephone (404) 977-5222.

National Agricultural Retailers Association, 1155 15th St. NW, Ste. 300, Washington DC 20005; telephone (202) 457-0825.

Golf Course Superintendents Association of America, 1421 Research Park Drive, Lawrence, KS 66049-3859; telephone (913) 832-4480.

ILLINOIS

STRUCTURAL

The Illinois Department of Public Health certifies applicators for structural pest control.

It has subcategories for the use of restricted pesticides: termites, bird control, fumigation, food processing, institutions, public health and wood processing.

But it also requires certification for use of general-use pesticides for those employed by commercial (for-hire) business.

Certification involves a test.

All commercial businesses must employ a certified technician, though non-commercial locations using only general-use pesticides do not have to have a certified technician.

Certification as a technician does not allow a technician to perform structural pest control for hire in Illinois unless employed by a licensed commercial business. If restricted pesticides are used in a non-commercial location (such as a food plant or apartment building), the location must registered with the department as well as employ a certified technician.

A $40 application fee is charged. The state provides a reference list of study materials to be obtained elsewhere.

Those taking subcategory exams to use restricted pesticides must have six months experience, 16 semester hours of college credit in related fields or a pest control course approved by the department.

Recertification is done every three years.

It is achieved by attending a recertification seminar during the three-year period and submitting a renewal fee. Retesting is not done unless the certification has been expired for more than one year.

AGRICULTURAL

The Illinois Department of Agriculture both certifies and licenses pesticide applicators.

Commercial applicators are certified for three years, the same amount of time the commercial applicator license lasts. The license costs $50 a year.

Licensees must prove evidence of financial responsibility and insurance.

Private pesticide applicators pay a $10 license fee.

The state also offers licenses under the following categories: public and commercial not-for-hire; licensed operator; pesticide dealers; and agrichemical facility.

An agrichemical facility is a site used for commercial purposes where bulk pesticides of more than 300 gallons or 300 pounds are stored or mixed more than 30 days per year. The state has a whole set of regulations regarding how agrichemical facilities must ensure they are built and maintained safely.

Private applicators are certified or recertified by doing one of the following: attending a training session conducted by Cooperative Extension and then passing a written closed book exam; or just successfully taking the exam.

Commercial applicators are also tested.

In 1993 Illinois passed a set of regulations aimed specifically at the lawn care industry. Under those rules, a sign must be placed on any lawn other than a golf course at the point or points of entry after pesticides have been applied. Golf courses must keep permanent signs up stating to visitors that pesticides are used from time to time and that further information is available from the superintendent.

Also, specific information must be available to customers of lawn care services and prior notification of application can be requested from either lawn care or golf course applicators.

All lawn care applicators for hire must be certified in Illinois.

The state also has rules specifying how wash water and rinsate should be collected and handled.

PESTICIDE POISONINGS WORLDWIDE

"It is estimated that there are three million severe pesticide poisonings each year worldwide, resulting in about 220,000 deaths, 99 percent of which occur in developing countries. There are 25 million episodes of pesticide intoxication annually in developing countries, while in the United States there are only 150,000 to 300,000 episodes of pesticide-related illness.

"Use of organophosphates is expanding, in spite of well-documented acute systemic effects resulting from exposure. Nerve damage occurs within hours of exposure, and paralysis may be evident 24 to 96 hours after acute poisoning."

Summary of paper in scientific journal The Lancet, July 27, 1991.

INDIANA

Both commercial and private applicators in Indiana must pass a core test, which is slightly different for each. Commercial applicators must also get a 75 percent score or better on another test in their category.

There are no fees for certification, but there is a $30 fee for getting a license, which is required of anyone applying any kind of pesticide for hire or restricted-use pesticides as a public or not-for-hire applicator.

The state offers a one-day training class for first-time certification, which costs $55.

Indiana has separate, more strict rules for two subcategories of commercial applicators: those working on turf and in structures.

To use restricted pesticides on turf or any pesticide on a for-hire basis on turf, a person must complete 90 days of active employment as a registered technician with a single employer. That requirement can be waived if the person goes ahead and becomes licensed as an applicator for hire, an applicator not for hire or a public applicator. But to do that on turf, there are additional rules: licensees must have been actively employed at least two years in the area or completed a formal post high school 2-year minimum turf program.

If the person decides to complete the 90-day registered technician program, he or she must then pass both a core and turf category certification exam and submit a license application and the fee.

Similarly, people wanting to be structural pest control applicators must have a year of active experience as a registered technician and work in for a licensed business before they can be licensed. To be a registered technician, one must take an open workbook exam, passing by 75 percent.

Before attempting to be licensed, the person must also submit case records for pesticide applications for which he/she has made inspections, prepared inspection reports, graphs, treatment proposals and treated the structure. For subterranean termites a total of 15 separate properties must be treated, and the rule specifies a variety of types of applications that must have been done. For wood destroying pest organisms, a total of three separate properties must be treated. Once all that is submitted, the applicant can go ahead and take the core exam and pest control exam for certification and licensing.

Certification lasts 5 years. To be recertified, applicators must either take the test again or accumulate the correct amount of continuing education credits.

Indiana also has rules on posting signs for outdoor lawn pesticide applications.

The sign, which must read "Lawn Care Application Keep Off the Grass," must be placed at a conspicuous point of access to the lawn and stay up during the day of application.

Customers must be provided with the name of the business, its telephone number, name of applicator, date and time of application, common name and end use concentration of each pesticide active ingredient, post application label safety precautions and instructions to the customer to remove the lawn marker.

IOWA

It's mandatory for both private and commercial applicators to take an exam every three years to apply restricted-use pesticides. It applies to all pesticides for commercial applicators.

Commercial applicators must take a 50-question core exam along with a 35-question category exam.

Certification costs $75 for commercial and $15 for private.

Iowa also has classifications for noncommercial applicators, who are persons who apply restricted-use pesticides on lands or property owned, rented, leased or controlled by the applicator or applicator's employer; certified handlers, who handle opened pesticide containers, repackage bulk pesticides or dispose of pesticide-related waste while supervised. There are tests for each category.

Iowa also has licenses for commercial applicator firms, noncommercial ones and public officials. The license costs $25, but is waived for public officials.

All applicators and handlers must also take six hours of continuing education each year.

Those certified applicators, along with handlers, recertify by retaking their exams or getting 2 hours of continuing education each of the three years between recertification.

The Iowa Pesticide Act requires that commercial, public and noncommercial pesticide applicators maintain records of each pesticide applied for three years after application.

In addition to buying an annual license, an applicant for a pesticide company license must file proof of financial responsibility either in the form of a certificate of liability insurance or a surety bond.

OTHER RULES

Iowa has several rather unique regulations that apply to pesticides.

One is the Iowa Bee Rule. It states that pesticide applicators contact the county office of the Agricultural Stabilization and Conservation Service to obtain all apiary locations with a two-mile radius of the field being sprayed. Beekeepers in the vicinity must then be notified of the application.

There are also special rules limiting use of atrazine to keep it out of the state's groundwater. In certain counties and townships designated as atrazine management areas, no more than 1.5 pounds of atrazine active ingredient per acre per calendar year may be applied. There are also 50-foot setbacks from bodies of water and wells.

Iowa also has a rule on secondary containment of pesticides on-site. This requires that nonmobile bulk pesticide containers be located within a watertight containment facility and any mixing or transfer of pesticides at a permanent storage and mixing site be done within a containment area. It spells out how a permanent site for storage or handling should be designed and paved to be water-tight.

KANSAS

Both private and commercial applicators must take a mandatory test to be certified in Kansas. But the private test is open-book, while the commercial one is closed-book.

Private applicators get their certification for five years and it costs $10. It is needed to apply restricted pesticides only.

Commercial applicators have an application fee of $35 per category along with an exam fee of $25 per category. Their certification is good for three years.

Commercial certification is broken down into 10 categories.

One of the requirements for a pesticide business license is that the owner or one or more applicator employees have commercial certification in each category in which the license application is made.

To renew certification, the commercial applicator must either take the test again or attend training approved by the Kansas State Board of Agriculture.

Pesticide users in categories for ornamental or turf pest control and wood destroying and structural pest control who use any type of pesticide have 90 days to become certified applicators or registered pest control technicians through approved training.

Kansas has reciprocal agreements with several Midwest states on certification.

It also has regulations on registering non-bulk pesticide storage and handling facilities.

There are also rules for pesticide businesses to have a surety bond or liability insurance; maintain records of applications; and register certain equipment.

Kansas has also created pesticide management areas that require special pesticide handling because of groundwater contamination problems.

KENTUCKY

Kentucky requires testing for commercial applicators, but not for private ones. Private applicators are trained using a video/slide program by the county extension agent. This training occurs every five years. There are also provisions for oral exams for illiterate private applicators.

Commercial applicators pay a $25 fee for certification in one of 12 categories (three more subcategories). It is good for five years.

The state also requires a pesticide applicator's license each year for a $25 fee, along with a $10 inspection fee for each aircraft licensed and for each piece of ground equipment to be licensed.

To be recertified, commercial applicators must take two approved training courses in five years.

Kentucky has additional regulations covering structural pest control and fumigation, with separate licenses for pest control applicators, pest control managers, fumigation applicators and fumigation managers. Anyone wanting a structural pest control applicator's license must have two years experience or a college degree in entomology. Managers must have one year's experience as a service technician or salesperson. These people must take a test and pay a $100 yearly fee, with higher fees for managers and licensed pest control operators.

The rules discuss minimum standards for termite treatment and control of wood-destroying fungi. There are also rules about fumigation, such as having fumigation crews have at least two people on them.

Kentucky prohibits local regulation of pesticides and sets out a statewide notification and posting rule. Those notifications cover applications on lawns.

FOR INFORMATION CALL
There is a toll-free number for questions about pesticides,
including health effects and how to deal with an emergency.
It's called the National Pesticide Telecommunications
Network and can be reached 24 hours a day at
1-800-858-7378. Its fax number is (806) 743-3094.

LOUISIANA

Private applicators are certified to use restricted-use pesticides by passing an examination.

Commercial applicators are certified in any of 10 categories. They must pass an examination.

There are additional requirements. People wanting to be certified in structure pest control on a non-fee basis in institutions or grain handling places must have two years' experience in the field. Anyone wanting to do mosquito control must have either a bachelor's degree with some entomology or four years experience working under the supervision of someone certified in mosquito control.

Louisiana also certifies people separately as pesticide salespersons and as agricultural consultants.

Certification is good for three years and costs $10 for private applicators and $15 for commercial.

Recertification can be done by attending a training meeting.

Owner-operators of pesticide control businesses must be licensed and show proof of financial responsibility.

OTHER RULES

Louisiana has a number of other regulations pertaining to pesticide use.

For instance, the state has its own list of restricted-use pesticides that are not restricted by EPA. There are 17 of them and they are banned during the summer months from use in 25 different parishes. However, waivers can be obtained.

Anyone wanting to apply pesticides aerially to rights-of-way for control of woody vegetation must notify the state ahead of time.

The state also must be notified of any bulk distribution facilities, and there are rules on storage of bulk pesticides. There are also lengthy requirements about disposal of containers.

MAINE

Maine does not participate in any form of reciprocity with other states. People seeking certification and licensing in Maine must complete written, closed-book exams with an 80 percent passing grade.

Farmers and others who need to apply restricted-use pesticides for purposes of producing an agricultural commodity on their own or leased lands must obtain a private applicator's license. To qualify, they must successfully complete both a core and a commodity specific exam. There is a $6 fee for a three-year license. During this period, they must earn six recertification credits to be eligible for license renewal.

Any person who uses pesticides in public places, on a for-hire basis or as a governmental employee must be licensed as a commercial applicator. There is a two-tier level of licensing and each company or agency must have at least one master to supervise their pesticide applications. Other employees may be licensed at the operator level and follow the directions of the master.

Commercial applicators must successfully complete a core exam and a category exam for each type of work they will perform. There are 26 different subcategory exams and there is a $10 fee for each category/subcategory exam taken.

There are prerequisites for a person seeking a master certification. They include the following: 1) worked under the supervision of a licensed commercial applicator for two years; 2) licensed as a commercial applicator for previous year; or 3) achieved a score of at least 70 in 25 credit hours of courses or received 40 hours of equivalent training.

A master candidate must also pass a closed-book exam on Maine law and regulations plus an oral exam covering: practical knowledge in ecological and environmental concerns; pesticide container and rinsate disposal; spill and accident mitigation; pesticide storage and on-site security; employee safety and training; potential chronic effects of exposure to pesticides; pesticide registration and special review; the potential for groundwater contamination; principles of pesticide drift and measures to reduce drift; protection of public health; minimizing public exposure; and use of non-pesticide control measures. In addition, the applicant must demonstrate the ability to interact with a concerned public.

Commercial applicators are certified for a five-year period during which they must accumulate between 8 and 15 recertification credits, depending on the category or subcategory. There is an annual $20 license fee, and incorporated firms and businesses with more than one applicator must also obtain a $100 annual spray contracting firm license. Commercial applicators must also verify insurance coverage and submit summary application reports on a quarterly basis.

OTHER RULES

Maine also has rules on restricted-use pesticide container disposal and storage, which includes a deposit system that encourages users to return properly rinsed containers for legal disposal.

The state also approved in 1988 special rules to minimize drift from powered equipment. Along with guidelines on how to avoid drift, there is a provision for the owner or occupant of a sensitive area to request to be notified when pesticides are to be applied within 500 feet of that area.

REGULATORY OFFICIAL ORGANIZATIONS

Association of American Pesticide Control Officials, Philip Gray, secretary, P.O. Box 1249, Hardwick, VT 05843; telephone (802) 472-6956.

National Association of State Departments of Agriculture, 1616 H St. NW, Washington DC 20006; telephone (202) 628-1566.

Association of Structural Pest Control Regulatory Officials, David E. Scott, pesticide administrator, Indiana State Chemist Office, Purdue University, 1154 Biochemistry Bldg, West Lafayette, IN 47907-1154; telephone (317) 494-1585.

Association of Official Analytical Chemists, 1111 N. 19th St., Ste. 210, Arlington, VA 22209; telephone (703) 522-3032.

Association of American Feed Control Officials, Barbara J. Sims, secretary, Office of the Texas State Chemist, P.O. Box 3160, College Station, TX 77841-3160; telephone (409) 845-1121.

Association of American Plant Food Control Officials, David L. Terry, secretary, Division of Regulatory Services, University of Kentucky, 102 Scovell Hall 00642, Lexington, KY 40546-0064; telephone (606) 257-2668.

Association of American Seed Control Officials, David Turner, secretary-treasurer, Oregon Department of Agriculture, 635 Capitol St. NE, Salem, OR 97310-0110; telephone (503) 378-3774.

MARYLAND

Both private and commercial applicators in Maryland must take a closed-book exam to be certified.

The private applicator must take both a core exam plus one specialty: field and forage crops, fruit, tobacco, vegetables, nursery and landscape ornamental, turf, greenhouse ornamental or wood preservation. Private applicators are considered individuals who use a restricted-use pesticide on their own or leased property to produce an agricultural commodity.

Commercial applicators must also have a year of pesticide experience for a licensed company or a degree in a biological field.

The fees are $7 for a three-year certification for a private applicator. Commercial ones must pay $100 for a business license, $50 for certification and $25 for each additional category they get certified in.

To be recertified, private applicators must attend a session in the last year of certification or take the test again.

Among the required topics being covered in the 1992-93 recertification sessions were: new laws and regulations, new recordkeeping requirements under the 1990 Farm Bill, worker safety standards, implementation of the Endangered Species Act and atrazine label changes and best management practices for Maryland.

Commercial applicators must attend one approved session each year for each category of certification.

There are 12 main categories with 14 subcategories.

Public agency applicators are exempt from fees.

OTHER RULES

Maryland also has regulations that require companies to give customers specific information approved by the Agriculture Department before making applications, including information on the pesticides being used.

There are sign posting requirements for exterior ornamental or turf applications.

Maryland also has a registry of pesticide-sensitive individuals. Any company making an exterior ornamental or turf application to a property adjacent to a person on the registry must notify the person in advance of the application.

MASSACHUSETTS

There are four types of licenses and certifications issued by Massachusetts' Department of Food and Agriculture: ·

Commercial Certification: Certification for applicators who use or supervise the use of any pesticide that is classified for restricted use for hire or compensation for any purpose or on any property other than as provided by the definition of "private applicator.''

Private Certification: Certification for applicators who use or supervise the use of restricted-use pesticides for the purpose of producing any agricultural commodity on property owned or rented by him or his employer or without compensation other than trading of personal services between producers of agricultural commodities on the land of another person.

Applicator License: Licensing for an individual to be present while restricted-use pesticides are being applied under the direct supervision of a certified applicator, or to use or supervise the use of general use pesticides for any purpose or on any land other than as for a private applicator.

Dealer License: Licensing by the Department to sell restricted-use or state limited-use pesticides.

There are about 5,500 individuals who hold licenses or certifications in the state and about 130 licensed dealers.

Both commercial and private applicators must pass an exam to be certified. It costs $25 per exam and $25 for a private certification, $50 for commercial certification. They last five years.

There are 20 categories of commercial applications and nine for privates, including cranberries and sod.

OTHER RULES

Rights of way: Any company or agency proposing to use a herbicide must use applicators who are certified or licensed by the department. Industries proposing to clear or maintain rights of way with herbicides submit a five-year vegetation management plan. A yearly operational plan is also submitted to the state and to municipalities in which proposed herbicide spraying is expected to take place.

Drinking water supply protection: As of May 1991, the state adopted regulations to prevent non-point source contamination of public drinking water supply wells from pesticides on the groundwater protection list. Pesticides on that list may not be applied in the primary recharge areas of public drinking water supply wells unless there is no viable alternative, an integrated pest management program is implemented and a pesticide management plan is approved for all applications.

Termite control: Applicators must notify customers in writing about the possible accidental contamination sometimes associated with subsurface applications and the steps to take to reduce the risk of such contamination. Customers must also receive a department-approved information sheet or a copy of the label, the material safety data sheet, the EPA fact sheet for the product intended for use and a copy of the regulations.

Agricultural aerial applications: Includes a permit system for fixed-wing applications, operational conditions to minimize off-target drift, setbacks from homes and surface water supplies and sign posting before applications are made.

Lawn care: Regulations that took effect in 1987 require certain information be provided to consumers about commercial applications to turf. They require a consumer information sheet be provided, pre-notification of the application to the customer, leaving information about the applicator and the product used at the property after application, and sign posting of treated lawns.

Exclusion provisions: These allow landowners to exclude their property from public area-wide applications of pesticides. Usually, the only pesticide application programs affected by the program are those for gypsy moth and mosquito control. Requests can be overridden by the commissioner of public health or by pest control officials who need to eradicate a recently introduced pest or one that is a threat to agriculture.

PESTICIDE, SCHMESTICIDE

There are pesticides, herbicides, termiticides, even slimicides. What do all these terms mean?

According to the federal law that regulates all these products, pesticide is the umbrella term that covers anything that kills pests, appropriately enough. That means if it kills a bug, a weed, or a slime it counts as a pesticide.

California's regulatory program prefers to call them "economic poisons."

MICHIGAN

In Michigan, the private applicator must take a private core exam, while the commercial applicator must take the commercial core exam and a category-specific exam for each category of application to be used.

They must attain a score of 70 percent or better. There is a fee of $10 for privates and $50 for commercials. Study manuals may be purchased to prepare for examination.

Recertification is required every three years. That takes place by recertification examination or attaining the necessary credits by attending Michigan Department of Agriculture approved seminar classes.

Any person applying general-use pesticides on the property of another, as a routine and schedule work assignment, is required to be either commercially certified or a registered technician.

The state tightened that requirement as of Jan. 1, 1992, when it changed the law so that only those applying general pesticides on their own premises or for a private agricultural purpose are exempt from the amendments.

That meant that employees, including janitors, custodians or maintenance personnel who may be applying pesticides as a routine and scheduled work assignment, became subject to the certification or registration law. Other examples include employees of schools, hospitals, nursing homes, public utility companies, apartments, campgrounds, cemeteries, golf courses, airports, railroads, state and local government and colleges.

To be a commercial registered technician one must be 18 years old, submit a $25 fee and pass the general standards exam and receive verifiable, category-specific training from a person who is an approved trainer. Approved trainers must have been a certified applicator for at least two years and have attended a one-day trainer program offered by Michigan State University.

Private certification includes private, aerial, structural fumigation, commodity fumigation, soil fumigation and greenhouse fumigation. Commercial certification has 27 categories.

Michigan requires anyone performing pesticide application for hire to hold a business license with at least one applicator having two years' previous experience.

MINNESOTA

Private applicators in Minnesota are certified by an open-book test. The certification costs $25 and lasts for three years. It is renewed by retesting and attendance at a workshop.

Commercial applicators take a closed-book, monitored exam. They pay $50 a year for a license and are recertified yearly for all categories but agricultural spraying, unless they attend a workshop.

Private certification is required for agricultural commodity growers who want to use restricted-use pesticides. Commercial certification is needed for anyone applying any pesticide for hire.

There are 16 categories of commercial applicator.

Minnesota also offers a noncommercial applicator license, designed basically for government applicators and other applicators who apply pesticides on their own premises or as custodians or food service workers.

There is also a structural pest control license. A person can be licensed as a master, journeyman, or fumigator. It involves taking a written or oral test and possibly a practical demonstration regarding structural pest control.

Minnesota's pesticide law also requires that training manuals and exams be continually updated and must address, among other things, groundwater contamination.

OTHER RULES

Minnesota also has provisions for the use of chemigation — applying pesticides through an irrigation system — that require a permit.

The state also provides for parameters for local ordinances on warning signs for turf areas, if local jurisdictions want to pass such laws, although they are not required to by the state.

The state law prohibits cities from enacting more restrictive pesticide application warning requirements.

MISSISSIPPI

Private applicators in Mississippi can be certified to use restricted-use pesticides by training or by a test, if they so choose.

Commercial applicators, however, must take a test on general standards as well as exam in the category they work in. There are 13 such categories.

Mississippi's Cooperative Extension Service provides study packets for each test at no cost.

Certification lasts for five years for private applicators and three years for commercial ones.

There is also a license requirement for anyone soliciting work or charging a fee for their services. To be licensed there are experience and/or education requirements to take the licensing exam.

Licensing is required for people doing the following work: entomological, plant pathology, horticulture or floriculture, tree surgery, weed control and soil classification.

INSECTS IN HISTORY

Bees have been the subject of conflicting superstition over the ages, according to the Dictionary of Superstitions.

As far back as the year 1050, people believed that if a person "dreams he sees bees fly into his house, that shall be the destruction of the house." The dictionary also notes a superstition from around 1850, when people believed that a bee flying into a house was a certain sign of death.

On the other hand, some people in the late 1800s believed that if they saw a "humble bee" fly inside it was a "sure sign of a coming visitor." Similar beliefs held that an insect should be left alone if it decides to enter the house; driving it outside would bring bad luck.

MISSOURI

In the "show me" state, applicators must show the state they can successfully pass state examinations to become certified as commercial, noncommercial or public operators.

Private applicators are required to attend a three- to four-hour certification training program for certification and licensure. The University of Missouri Extension Service conducts certification training. There is no exam for private applicators, and the certification lasts five years.

The other three categories must recertify every three years.

Their initial certification costs $50 for a commercial applicator, $25 each for noncommercial, pesticide technician and pesticide dealer, and no charge for public operators.

There are additional requirements for being licensed in structural pest control. Applicants must have a minimum of an AA degree in agriculture, biology, chemistry or entomology; or a year's experience.

PESTICIDE RESISTANCE

According to the Worldwide Insecticide Resistance Database at the University of California at Riverside, over the last 40 years nearly 600 species of insects have become resistant to major classes of insecticides.

From the 1993 Earth Journal Environmental Almanac and Resource Directory

MONTANA

The Montana Pesticides Act requires that persons needing to be licensed are:

1. Pesticide dealer, any person who sells pesticides within the state, except those used for home, yard and garden use.

2. Pesticide applicators:

—commercial applicator, a person, who by contract or for hire, applies pesticides to land, plants, seed, animals or water.

—non-commercial applicator, a person who can't be classified private, commercial, government or public utility but desires the use of pesticides.

—public utility applicator.

—governmental applicator.

3. Private applicator, any person who applies restricted pesticides to his own crops or lands. The private license costs $50.

Dealers get a license by passing a written exam with a 75 percent or better score and submitting a license fee of $75.

Commercial applicators must pass written exams with 70 percent or better score for general use products or 80 percent or better score for restricted-use products; meet liability insurance requirements; and pay a $75 license fee.

Government applicators must get the same minimum scores as commercial and pay a $75 licensing fee.

Licenses are good for one year.

Certification is good for five years for private applicators and four years for commercial and government.

To be recertified a private applicator must take six hours of coursework. Commercial and government applicators must take 12.

Montana also has special rules on aquatic herbicides, M-44 and livestock protection collars.

NEBRASKA

The U.S. Environmental Protection Agency handles certification of private and commercial applicators in Nebraska. That occurred after a dispute in 1977 between the state and EPA over whether the state's plan for certification and training would be adequate. EPA decided it would not be, and, as required under FIFRA, took on the responsibility for certifying an estimated 7,500 commercial and 37,400 private applicators who use restricted-use materials.

Commercial applicators are required to pass a written exam by 70 percent to be certified in one of 13 categories.

Because FIFRA does not allow a mandatory test for private applicators (other states do it because they can be more stringent than the federal government under FIFRA), private applicators in Nebraska can do one of several things if they want to use restricted-use pesticides: complete a training course, take an exam; complete a self-study program; submit evidence of certification as a private applicator in another state.

During 1994 Nebraska will take over the applicator certification program from the federal government.

EPA IS EVERYWHERE

EPA pesticide regulations cover:

—About 30 major pesticide producers and another 100 smaller producers

—3,300 formulators

—29,000 distributors and other establishments

—40,000 commercial pest control firms

—About 1 million farms

—Several million industry and government users

—About 90 million households

From EPA's Pesticide Programs, an agency publication from May 1991

NEVADA

Both private and commercial applicators must take an exam to be certified. Certification costs $5 and lasts four years.

There are 12 categories for commercial applicators.

There are also three categories and 11 subcategories for private applicators.

Nevada issues different types of certificates:

—a general certificate authorizing the application or supervision of the application of restricted-use pesticides for the category issued.

—a limited certificate authorizing the application or supervision of the application of restricted-use pesticides, on a specific host, for a specific pest, or performing specific application procedures.

—a single use certificate authorizing the application or supervision of the application of a single restricted-use pesticide on a one-time basis, for an emergency measure.

The passing score for an exam is 60 percent.

Nevada also requires a license to engage in pest control or in pest control activities concerning wood-destroying pests or organisms.

NEW HAMPSHIRE

Private applicators in this state wanting to use restricted-use pesticides must pass an exam.

Commercial applicators must pass an exam in one of 11 categories.

To be recertified after five years applicators may either take related training coursework during that time or retake the examination. Applicators must have a total of four credit units of general information and eight credit units of commodity-specific information for each commodity they hold.

New Hampshire also gives certification at the operational level, requiring passage of a written general exam. Applicants must be employed by a pest control firm.

The state gives out supervisory level certificates to applicators who have been licensed at the operational level for five growing seasons, are graduates of a college or university in an insect or growing-oriented major, or are graduates of a two-year agricultural or technical school and have been licensed at the operational level for three growing seasons under the supervision of a person holding a supervisory license. They must also pass both written and oral examinations.

At least one employee at the supervisory or managerial level of each business entity applying pesticides commercially is required to hold a supervisory level certificate of registration. At least one member of each crew shall be registered at either the operational or supervisory level.

Pesticide dealers must get a license by paying a $20 license fee. They must also take a test. Each employee or agent of a pesticide dealer who sells or recommends restricted-use pesticides must obtain a pesticide dealer license and take an exam.

Every business entity engaged in commercial application of pesticides must have a certificate of registration.

The state has additional restrictions on applications to rights-of-way and woodland areas, as well as near wellheads.

Landowners near the right-of-way spraying must be notified.

The state also requires notification prior to application of pesticides to turf areas. Signs must be posted when applications are made to multifamily dwellings or to public and recreational properties.

New Hampshire also has its own list of restricted materials.

NEW JERSEY

Private applicators in New Jersey must take a basic core certification exam to apply restricted-use pesticides. Once that is done, a second, mail-in exam must be returned within four months.

Certification for private applicators lasts five years.

They can be recertified after that time by taking the two tests again. However, the preferred way is to accumulate units of recertification credit over the five-year period by attending courses, seminars, meetings etc. which deal with some aspect of pest control or pesticides.

A private applicator would have to accumulate 24 units over the five years. Eight of the units are for core credit 16 for private part 2 credit. Core credit is given for instruction covering material like that covered in the core exam, such as pesticide regulations, the safe handling of pesticides and understanding pesticide labeling. Private Part 2 credit is given for instruction covering material such as pest biology and identification, specific pesticides and various pest control methods.

Commercial applicators must take two exams, both of which are monitored. The second test is in one of 11 basic categories, many of which have subcategories.

Training manuals for each exam are available from county extension offices.

Commercial applicators also get recertified after five years through either retaking the tests or attending courses. To do so, they must accumulate eight units of core recertification credit and 16 units of category credit for each category they are certified in.

New Jersey also registers commercial pesticide operators. A commercial pesticide operator is any person who handles or applies pesticides, by equipment other than aerial, under the supervision of a commercial certified pesticide applicator. Operators not required to register are those supervised by a commercial pesticide applicator who is always physically present and in line-of-sight of the one being supervised when pesticides are handled or applied.

Private pesticide operators must also register with the state.

This does not require passing a test. Instead, the operator must obtain instruction in the proper use and application of pesticides.

Then they fill out a verification form. A certified applicator is responsible for making sure the training has been undertaken according to state guidelines.

There is another state requirement for registering a pesticide applicator business, considered a business or person who either wholly or in part holds himself out for hire to apply pesticides, such as exterminators, landscapers, tree services and pet groomers.

Businesses must register each location where they have an office and every name they do business under. Such businesses must also verify insurance coverage, with the amount being higher for those engaged in fumigation work.

Pesticide dealers in New Jersey must also register. There is a dealer certification examination to be passed. There is a dealer business registration, too.

It costs $10 for each exam, and licenses are $75 for commercial applicators,

$30 for commercial operators, no cost for private applicators and operators, $150 for pesticide applicator businesses, $75 for certified dealers and $150 for dealer businesses.

GOLF COURSE CERTIFICATION

Unlike many other pesticide users, golf course superintendents have the opportunity to have their training and expertise certified by an entity other than the federal government. An industry group, the Golf Course Superintendents Association, has a comprehensive series of continuing education courses and professional certification.

The organization says that its certification program is popular within the industry because it is a good way for superintendents to prove their worth to prospective employers.

Over time, the certification program is tightening up its standards for superintendents to even apply for the program. As of July 1, 1994, applicants must have an associate's degree (60 semester hours or completion of the group's Division I continuing education coursework. In 1999 that entry requirement goes up to 90 semester hours of college or Division I and half of Division II; and in 2004 applicants must have a bachelor's degree or Division I and Division II completed.

The GCSA offers continuing education programs for basic turfgrass management, advanced turfgrass management, executive development, technical training and correspondence courses.

The dozens of courses include disease identification, protection of public water resources, golf course design and media relations.

The GCSA can be reached in Lawrence, Kansas at (913) 832-4444.

NEW MEXICO

A general, core exam is mandatory in New Mexico for both private and commercial applicators.

Commercial applicators must also take two exams — one general and another on laws and regulations — as well as one in whatever category they work in.

Private applicators pay nothing for the test, but pay $5 to be licensed as such. It lasts five years.

For commercial applicators, it costs $50 to be certified and licensed. Before they can take the exams, applicants must have 20 college credit hours and a year of pesticide application experience OR two years of pesticide application experience.

They must also prove financial responsibility.

Any person who uses any pesticide as an employee of a commercial applicator must be licensed as an operator or serviceman. They must pass the appropriate operator examination, pay an exam fee of $5 and an operator license fee of $25.

NEW YORK

The state has six categories for private applicators, who must be certified if they use restricted-use pesticides. To do so, the private applicator must pass an exam. Certification is then good for three years.

To recertify, the private applicator must take at least three refresher courses (10 credits) and recertify every six years.

Commercial applicators must take a core exam along with a test in whatever category or categories they will work in. They must renew every three years.

To be recertified after six years they must attend at least three refresher training courses during that time. There are a specific number of credits required for each category or subcategory.

Each business providing the services of commercial application of pesticides must register each year with the department and pay a $50 fee. Such businesses must also provide the state with proof of insurance. A commercial certified applicator must be employed at each business location.

Other rules require specific disclosure of information to people hiring pesticide applicators for lawn, tree and shrub applications involving signage. Prior notification is required before all pesticide applications.

There is also a special rule for protection of grape-growing areas.

FATAL FARMING

Farming is one of the more dangerous occupations, with 600 work-related deaths in 1992 to farm residents and 230,000 disabling injuries, the National Safety Council reports.

Tractor accidents, particularly overturning tractors, are big factors.

Agriculture work includes forestry and fishing.

NORTH CAROLINA

Dealers who sell pesticides classified for restricted-use by the EPA or North Carolina Pesticide Board to the end user must pass a licensing test based on the information contained in training manuals. A licensed dealer is required at each sales outlet. The annual licensing fee is $30. Dealers who sell pesticides other than those classified for restricted-use do not have to be licensed.

Commercial pesticide applicators, except those doing structural work, must pass a licensing test based on training manuals put out by the state. They are tested in one or more specialty areas. The state lists 13. This includes, but is not limited to, private golf course operators, wood treaters, seed treaters, and research personnel.

The annual licensing fee for a commercial ground or aerial applicators, regardless of the number of specialty areas needed, is $30. Licenses expire Dec. 31 of each year.

Recertification by means of continuing education or reexamination is required every five years for ground applicators and every two years for aerial applicator. Recertification hours for a five-year period range from 10 for agricultural pest plant specialties to three hours for seed treatment.

Public operators are people who apply or supervise the application of pesticides in their jobs for town, city, or other governmental agencies. They take the same tests as commercial, non-structural applicators. No fee is required for their license.

Public utility employees who apply or supervise the application of pesticides are considered commercial applicators and pay an annual licensing fee of $30. Public utility employees and public operators are also subject to recertification requirements.

Consultants are people engaged in the business of giving pest control advice for a fee. Normally, they neither sell nor apply pesticides. Consultants must demonstrate expertise in each specialty area where they provide a service. They pay an annual licensing fee of $30. They must also be recertified every five years.

North Carolina recordkeeping requirements for commercial applicators keeping track of restricted-use pesticide applications are stricter than federal requirements. Under the federal law it's two years, while under the North Carolina Pesticide Law of 1971 those records must be on hand for the past three years.

Private applicators, usually farmers or ranchers, who buy and use restricted pesticides on their own or rented land must be certified. They can do so by participating in a four-hour extension training program, completing a programmed instruction exercise or taking a test given by the state Department of Agriculture.

There is no fee for private applicator training but there is a $6 fee for certification. They must be recertified every three years by retraining or reexamination.

Certified structural pest control applicators obtain certification in one of three phases: household pests, wood destroying organisms or fumigation. They are tested, at $10 per phase. An annual renewal fee of $30 is charged.

People in charge of a business engaged in structural pest control for the general public must be licensed by the Structural Pest Control Division of the Department of Agriculture. A prospective licensee must have prior certification in the phase or phases in which licensing is desired.

Structural pest control applicators and operators must be recertified by taking a test or the following hours of coursework: five hours per five-year period in one phase, seven hours for two phases and nine hours for three phases.

NORTH DAKOTA

Private and commercial applicators must pass open book tests to become certified. The private applicator takes a 30-question test and is certified for five years. The commercial applicator takes a long test and is certified for three years.

Fees are $10 for private and for commercial it's $30 for certification in one category, $40 for two categories and $50 for three or more categories.

Commercial applicators are broken down into 12 categories, while there are two categories for private applicators.

FUN FIFRA FACTS

The Federal Insecticide, Fungicide and Rodenticide Act is one of many environmental laws the Environmental Protection Agency carries out. But it is unique in one major respect: instead of EPA simply acting when it sees a significant risk to the environment, the law requires the agency to balance the risks and benefits of any decision.

While pesticide users are pleased with that fact, it complicates decisionmaking quite a bit.

For instance, EPA has to have good information on what the benefits of a pesticide are to fill in that side of the equation. But often, because farm groups have resisted mandatory recordkeeping as cumbersome, there is little or poor information on how much of a given pesticide is actually used in any given situation. That's why there's a push to get that use information from growers, pest control companies and other users. California already does it by requiring anyone using any commercial pesticide to check in with the local agricultural commissioner and fill out use reporting forms each month. The idea is likely to spread in the future.

OHIO

Both private and commercial applicators must pass a written test to be certified in Ohio.

Certification lasts for three years and costs between $20 and $100.

There are 30 different categories of commercial applicator and 13 for private.

Commercial applicators must also fill out a financial responsibility statement, while private applicators do not.

Commercial applicators must get a "custom applicator license" each year.

There is also a license category for custom operators and public operators, as well as limited commercial applicators.

Ohio also has rules for notice of lawn pesticide application.

When applying a lawn pesticide to residential lawns in any city or subdivided area of a township, an applicator must leave in writing with the person whose property was applied information about the chemical itself, the maximum concentration applied, any special instructions, the company's name and phone number and the date and time of application.

Neighbors can ask for advance notice of the lawn application.

OKLAHOMA

Oklahoma certifies pesticide applicators in three classifications: private applicator, certified applicator and service technician.

Private applicators take an open book exam for certification, which lasts for five years.

Commercial applicators must pass a closed-book test and pay a $20 fee for each category they test in. There are 20 categories.

Commercial applicators must be certified to use all pesticides, not just those listed as restricted-use.

Anyone wanting to be certified for general pest control, structural pest control, fumigation or food processing must also take a practical exam along with the general core exam and a category exam.

There is a separate test for service technicians.

Oklahoma has licensing. It is required of anyone to act, operate, do business or advertise commercially or noncommercially.

Under that rule, a commercial applicator is any person who engages in commercial application or commercial employment of devices. They must provide proof of financial responsibility.

A noncommercial applicator, who must also be licensed, is any person besides a commercial or private applicator who uses or supervises the use of a restricted-use pesticide under the supervision of a person who owns or manages the property and who has been certified. This category includes government employees.

There is also a licensing category for consultants, who are people who make pesticide recommendations for hire or compensation but do not purchase or apply the pesticide.

Fees for licensing are $50 for each category but not more than $250 total annually. Noncommercial licensees pay $20. Consultants pay $50 for each category and not more than $250 annually. No fees are charged to governmental agencies or their employees.

To be recertified, certified applicators either take the written test for each category or collect enough continuing education units to recertify in a category. Fee for recertification is $20.

Service technicians and private applicators can recertify only through written examination.

OTHER RULES

Oklahoma also has rules on recordkeeping, pesticide storage and reporting of spills.

Commercial and noncommercial pesticide applicators must keep accurate records on each application for two years. Additional information must be kept on hand by structural pest control applicators.

The guidelines say storage areas should have drainage systems to catch runoff water, be able to be securely locked, be identified with warning signs, among other things.

Uncontained spills of more than 10 gallons liquid, 25 pounds dry weight or 50 gallon of mixed pesticide must be reported within 24 hours to the state.

The state also has minimum standards for termite work on various types of structures.

Oklahoma regulations also stipulate certain areas of certain counties where "hormone-type" pesticides such as 2,4-D, dicamba and picloram may not be used between certain dates.

HERE COMES THE SUN

Skin cancer is a special problem for people who spend a lot of time working outdoors, such as farmers. The incidence of the disease is increasing by 3.4 percent each year with an estimated 700,000 new cases of cancer diagnosed in 1993.

The National Farm Medicine Center has a screening program to detect cancer early, when it is most easy to cure.
Of those participating in 1991, 51.9 percent had some type of skin damage. Of those who obtained follow-up cases, 30 percent were diagnosed with skin cancer and treated. Fewer than one-fourth used sun protection measures on a regular basis, even though most participants were aware of the risk.

Accident Facts, 1993 Edition, National Safety Council

OREGON

Oregon's Department of Agriculture has a number of different license types for pesticide applicators, some of which require passage of an exam and others that don't.

Private pesticide applicator: A person who uses restricted-use pesticides on land in agricultural production owned, leased or rented by him or his immediate employer. Must pass a private applicator examination and pay a $25 licensing fee.

Commercial pesticide operator: A business that applies any pesticide for others as a commercial activity. Must complete financial responsibility insurance certificate, pay a licensing fee and hold commercial applicator license.

Commercial pesticide applicator: Person employed by a licensed commercial operator, in direct charge of or supervising the application of pesticides or a person other than a private or public applicator who applies restricted-use pesticides. Must pass laws and safety examination and category examination and pay a $15 fee for the test plus $7.50 for each additional category.

Public pesticide applicator: A person employed by a governmental agency or its subdivision or public utility in direct charge of or supervising the application of restricted-use pesticides or any other pesticide applied with machine powered equipment. Must pass laws and safety exam and category exams, plus pay fees.

Directly supervised commercial pesticide trainee: A person working under the direct supervision of a properly licensed commercial applicator (on-site supervision not required). Must pass a trainee examination and pay a fee. Can only be renewed once.

Immediately supervised commercial pesticide trainee: Person working under the immediate, on-site supervision of a licensed commercial applicator. No exam, pays licensing fee and can be renewed indefinitely.

Directly supervised public pesticide trainee: A person working under the direct supervision of a licensed public applicator, with no on-site supervision necessary. Requires passing score on trainee exam and license fee. Can only be renewed once.

Immediately supervised public pesticide trainee: Person working under the immediate, on-site supervision of a licensed public applicator.

Pesticide consultant: Person who offers or provides technical advice to the users of restricted-use pesticides. Requires passing score on pesticide consultant exam and licensing fee.

Pesticide dealer: Pesticide sales outlet that sells restricted pesticides. Pays licensing fee, no exam.

Licensed applicators who need to be recertified can do so by either taking the test(s) again or completing a certain number of hours of training. Recertification is necessary after five years.

Private applicators need to have four hours each of Core A and Core B training and another eight hours of other training.

Pesticide consultants, public pesticide applicators and commercial pesticide applicators must earn 40 hours of accredited training.

Oregon also has rules on keeping records of pesticide application. They apply to pesticide operators, public pesticide operators and commercial pesticide applicators not employed by operators. Those records must be kept for three years.

PUBLIC CONCERN

In a 1993 survey by the consumer organization Public Voice for Food and Health Policy, 92 percent of those surveyed expressed concern about the health effects on young children of chemicals used to grow food. Sixty-eight percent were very concerned.

By comparison, the same survey showed just 55 percent of those questioned were very concerned about health problems from secondhand cigarette smoke, 47 percent about industrial and vehicular air pollution and 61 percent about the risk of severe food poisoning from bacteria in meat.

PENNSYLVANIA

Pennsylvania's Department of Agriculture regulates pesticide dealers, pest management consultants, pesticide application businesses, commercial and public applicators, pesticide application technicians, and private applicators.

Pesticide dealers must keep records on all restricted-use pesticides for three years. They must also get a license, for a fee of $10.

Pest management consultants must pass a written exam to receive a license, with a fee of $25.

Pesticide application businesses must be licensed, and it costs $25 a year. The license is issued in any of a number of categories. The business must prove financial responsibility and may not allow someone to make an application in a category for which they are not certified as an applicator or trained and registered as a technician.

Commercial and public applicators must be certified. They are given a written examination covering knowledge and competence in the law, safety and comprehension of pesticide labels. One part of the exam is a core area with general questions and the second covers specific categories the individual works in.

Commercial applicators must get a certificate, which costs $30 a year.

Public applicators must get a certificate, which costs $10 every three years.

The commercial/public applicators core examination is $50. The commercial/public applicator category examinations are $10 each.

To be recertified, commercial and public applicators must provide evidence of having received current update training in technology relating to pesticides in the specific categories in which the applicator is certified.

Pesticide application technicians must be trained in the law, spill handling, human health and environmental effects, as well as on-the-job training. Each technician is registered yearly for a $20 fee.

Private applicators are certified with a written examination, which can be taken again is a passing grade has not been achieved. Every three years the private applicator must have accumulated credits toward recertification based on attendance at meetings or other appropriate training approved by the Department of Agriculture.

There are special permits for private applicators who wish to use fumigants. They must already be certified and also take a test on the specific fumigant they want to use.

The examination required of private applicators is not charged for.

Pesticide management consultants must also take an exam, for which there is no charge unless an exam is requested on other than regularly scheduled exam date.

There are 25 categories for certified and public applicators to be certified in.

OTHER RULES
Pennsylvania also has a pesticide hypersensitivity registry for people who

have been verified by a physician to be hypersensitive to pesticides. Commercial and public pesticide applicators are provided the registry list and are asked to contact anyone on the list whose property is within 500 feet of a pesticide application site and inform them prior to application.

NOBODY LIKES A PEST

The English language has an apparent obsession with the concept of pests, or bothersome things or annoying, vexing and bedeviling things. In fact, Roget's Thesaurus lists dozens of synonyms for "pest."

So if you're tired of doing "pest control," try using one of these words instead:

Annoyance, vexation, bothersome, exasperation, aggravation, nuisance, bother, trouble, problem, difficulty, trial, bore, drag, downer, worry, bad news, headache, pain in the neck, harassment, molestation, persecution, dogging, bounding, harryment, devilment, irritation, exacerbation, salt in the wound, embitterment, fret, gall, chafe, or pea in the shoe.

And if those bugs have really got you down, try wretchedness, despair, bitterness, infelicity, misery, anguish, agony, woe, bale, melancholy, depression, sadness, grief, torment, torture, purgatory, nightmare, or even hell on earth.

RHODE ISLAND

Both private and commercial applicators in Rhode Island must take an exam as well as attend training sessions to be certified.

Private applicators go to a two-day training program and take an exam in the appropriate commodity category they work in.

Commercial applicators go to two-day core training and get additional training in each category. They also take exams.

Core training costs $50; category training costs $40.

Certification of private applicators costs $20. Commercial applicators pay $45.

Certification lasts five years for both.

Commercial applicators are certified in 10 different categories.

To be recertified they must attend at least one training course.

The state also issues commercial applicator licenses to anyone who uses general use pesticides as a commercial applicator. There is a $30 fee for the license. Applicants must show proof of financial responsibility.

Dealers must also be licensed if they sell restricted-use pesticides.

The state also has other rules on keeping applications away from wells and away from watersheds for public water supplies.

SOUTH CAROLINA

Commercial applicators in South Carolina obtain study material from the Department of Fertilizer and Pesticide Control and then pass a core, general exam and a specific category exam.

Private applicators do not have to pass an exam. They are trained by cooperative extension service local offices. They are then given an application for a private applicator's license on the day of training.

The state also has a category for noncommercial applicators, which includes government employees.

The state has 12 categories for commercial applicators to be certified under.

Certification lasts for five years. After that, applicators can be recertified by either taking an exam or accumulating continuing certification hours of training.

Commercial and noncommercial applicators must accumulate 10 such hours and private applicators must accumulate five hours.

South Carolina also has a law devoted to chemigation.

ACCIDENTAL POISONING

If someone is accidentally poisoned by a pesticide, five steps should be taken immediately:

1. Move the patient away from the pesticide and remove the patient's contaminated clothing; wash patient off.

2. Start first aid at once. If breathing is weak, give artificial respiration.

3. Call a physician; do not stop first aid.

4. Keep patient as quiet, warm and comfortable as possible.

5. Rush patient to hospital or other medical help.

Remember to save the label to show to the hospital or physician that the patient sees.

The basic rule is to stay calm and remember that the most serious effects of most pesticides are not instantaneous, so there is time to act.

From Colorado Department of Agriculture study guide

SOUTH DAKOTA

Certification required for an applicator's license is maintained by taking and passing a written open book examination every two years following initial certification by a written open book examination.

In lieu of an examination, attendance at a certification shortcourse held under the direction of the department can qualify the licensed applicator for maintenance of certification.

Pesticide dealers must be licensed by paying a fee and taking an exam.

Private applicators are certified by completing either a training course approved by the department or a home study course approved by the department.

There are also rules addressing secondary containment of bulk pesticide storage areas.

Beginning Feb. 1, 1994, all applicators who conduct operational area activities must use procedures to minimize and mitigate the adverse effects of discharges on the environment by keeping a written plan and training employees. As of that date operation areas using containment under the rules must be registered with the state.

TENNESSEE

Tennessee does not test private applicators.

Commercial applicators must be certified if they use restricted-use pesticides. All pest control technicians must be certified unless they work under the direct supervision of a certified person.

To become commercially certified an applicator must pass one exam that is a combination core and category exam. Study materials are available from the University of Tennessee Agricultural Extension Service for a small fee.

In most cases Tennessee will reciprocate certification with other states.

Recertification points are accumulated to recertify after five years.

The required number of points ranges from 12 to 28, depending on the category. Tennessee has 13 categories of commercial applicators.

Anyone certified after April 21, 1995, must accumulate between seven and 17 points.

Each separate office offering application of pesticides for a fee must have at least one licensee for each category of service. License exam applicants, usually the owner or supervisor of technicians, must qualify in one of three ways: have two years verifiable working experience under a licensee in that category; possess a four-year college degree with a major in a related field; or hold a similar license in another state.

Passing the exam entitles one to apply for a license. There is a fee of $20 per year per category, except for consultants, who must pay $250 per year. Passing a license category automatically certifies a person in the corresponding commercial certification category.

Each business offering the application of pesticides for a fee must also hold a pest control charter. There is an annual fee of $200. The registration for non-clerical employees such as solicitors and technicians is $20 per person per year.

TEXAS

DEPARTMENT OF AGRICULTURE

Private applicators must take a training course and pass a test. Commercial applicators do not have to take training but they have to pass a minimum of three tests.

Certification lasts for five years for private applicators and they pay a fee of $50. Commercial and noncommercial applicators pay $150 per year and $100 per year, respectively, plus $20 per category test. Their certifications last for three years.

There are 23 categories for commercial and non-commercial applicators to be certified in.

Applicators are licensed only if they use restricted-use or state limited use pesticides. The agriculture department does not require a license for application of general use pesticides.

Commercial and noncommercial applicators licensed by the state must earn a minimum of five hours each year for annual renewal and 15 hours every three years for recertification. They must also obtain 2 hours in laws and regulations and 2 hours in integrated pest management.

Certified private applicators may satisfy recertification with 15 hours of continuing education, plus 2 hours in laws and regulations and 2 hours in integrated pest management.

STRUCTURAL PEST CONTROL BOARD

Any person engaged in the structural pest control business must secure a business license from the board for each business location. They must then designate a responsible certified commercial applicator.

The board certifies the following categories:

Certified commercial applicator: The person responsible to provide training and direct supervision for pest inspections, identifications and control measures of a licensed business must be a certified commercial applicator.

Certified noncommercial applicator: The person who as an employee is responsible for providing pest control services to a governmental entity, apartment building, day care center, hospital, nursing home, hotel, motel, lodge, warehouse, food processing establishment, school or education institution.

Technician: Individuals who perform pest control services under the direct supervision of a certified applicator.

Technician-apprentice: Individuals who perform pest control services under the direct supervision of a certified commercial applicator must obtain a technician-apprentice license prior to obtaining a technician license.

To qualify to take the board test for obtaining a certified commercial applicators license the applicant must have verifiable employment in the pest control industry under the supervision of a licensed certified commercial applicator for at least 12 months out of the past 24 months and must have possessed a technician license for at least six months.

The board charges $66 for an annual license fee and $30 per examination.

Certification lasts for three years.

Categories are pest control, termite control, lawn and ornamental pest control, weed control, structural fumigation, commodity fumigation and wood preservation.

The state is beginning testing requirements for technicians.

INTEGRATED PEST MANAGEMENT

The effort to get away from pesticide use is mainly focused on an alternative method called Integrated Pest Management. That basically means using pesticides as a last resort and spending more time considering other alternatives. It can be used in a farm field, on a lawn, in a home or anywhere else there are pests.

According to a description in a Colorado Department of Agriculture publication, the following are the components of an IPM program:

—Identification of pests and natural enemies.

—Monitoring and recordkeeping system for regular recorded observations of pest and natural enemy populations along with other variables such as weather.

—A determination of the economic or aesthetic injury level, meaning, Does the potential harm warrant doing something?

—A determination of action levels - the pest population size from which it can be predicted that injury levels will be reached within a certain time if no action is taken.

—Selection of strategies and mixes of tactics that are least disruptive to natural controls and least hazardous to human health and the environment.

—An evaluation system to determine the outcome of any actions.

For more information on IPM contact a local Cooperative Extension agent or the Bio-Integral Resource Center, P.O. Box 7414, Berkeley, CA 94707.

UTAH

Private applicators do not have to take a test to be certified to use restricted-use pesticides in Utah. They take coursework instead.

Commercial applicators, considered anyone using pesticide for hire or compensation, are certified in one of 12 categories.

There is also certification for noncommercial applicators, who is any employee who uses a restricted-use pesticide and doesn't qualify as a private applicator or require a commercial applicator license.

Commercial applicators must get a license. To do so, they must pass by 70 percent a general exam as well one in their category.

Certification lasts for three years.

VERMONT

Commercial applicators must be certified before they may apply any pesticide. Private applicators must be certified to use restricted-use pesticides.

Private and commercial applicators are required to take and pass examinations in order to become certified.

All examinations are closed book exams. 75 percent is the passing score.

Commercial applicators take a CORE and category exams. Private applicators take the private applicator exam.

There is no fee for taking the exams. There is a $20 fee for certification in each commercial category. There is no fee for certification as a private applicator.

Certification is valid for five years.

All people selling pesticides must have a dealer's license.

All business enterprises applying pesticides commercially must be licensed. This is a $40 annual fee.

Warning signs must be posted after all turfgrass and landscape pesticide applications.

All golf courses in Vermont must obtain a permit before using pesticides.

No one may apply an herbicide to a right-of-way without first obtaining a permit from the Commissioner of Agriculture.

FUN WATER FACTS
The Earth is three-quarters covered with water but 97 percent of it is salty ocean and 2 percent frozen in ice. The other 1 percent is surface and groundwater.

VIRGINIA

Virginia has a mandatory exam for private applicators, and splits them up into seven categories: agricultural, nursery/greenhouse, fumigation, chemigation, aerial, limited certificate (single product/single use) and single product certification.

There is no fee for private applicators to take the exam. To recertify, they may either take courses or retake the exam.

As for commercial applicators, Virginia has 12 categories and 13 subcategories in which to be certified. There is a $35 per year fee and certification is done every two years.

To take the exam a person must have a registered technician certificate with one year's work experience as one; or a statement from the employer that the person has performed duties similar to those of a registered technician for at least a year and have the equivalent practical knowledge of proper pesticide use.

Virginia also requires a $50 business license for any person or business that sells, stores, distributes, mixes, applies or recommends pesticides for use for compensation.

Virginia also offers a registered technician certification. It is designed for an individual who renders services similar to those of a certified commercial applicator, but who has not completed all the training or time in service requirements to be eligible for examination for certification as a commercial applicator. Registered technicians may use general-use pesticides but may apply restricted-use pesticides only when under the direct supervision of a certified commercial applicator.

To get the registration, a person must receive on-the-job training under the direct, on-site supervision of a certified commercial applicator for at least 40 hours during the six month period previous to applying, along with board approved training; and pass a registered technician exam. It costs $15.

The state has also created a commercial applicator not-for-hire designation for commercial applicators who use or supervise the use of pesticides as part of job duties only on property owned or leased by him or his employer. Among the people who would fall under this category are golf course groundskeepers, apartment complex custodians and government employees.

Commercial applicator not-for-hire exempts one from the business license requirements, but there are requirements to show evidence of financial responsibility and recordkeeping.

WASHINGTON

Washington's Department of Agriculture has the following categories of applicator:

Commercial Applicator, annual license fee $142: A person engaged in the business of applying pesticides to the land/property of another. This land can either be publicly or privately owned. Prior to license issuance, a financial responsibility insurance certificate must be filed with the department by the insuring company.

Commercial Operators, annual license fee $39: A person employed by a department-licensed commercial applicator to apply pesticides to the land of another. This land can either be publicly or privately owned.

Commercial Pest Control Consultant, annual license fee $36: A person who sells or offers pesticides for sale at other than the licensed pesticide dealer outlet from which they are employed.

In addition, commercial consultants may offer or supply technical advice, supervision, aid or make recommendations to the users of non-home and garden pesticides. They may also perform wood destroying organism inspections. Licensed and employed commercial applicators and commercial operators may act as commercial consultants without acquiring the consultant's license.

Dealer Manager, annual license fee $21: A person who supervises the distribution of pesticides (other than home and garden products) from a licensed pesticide dealer outlet.

Private Applicator, annual license fee $23: A person who applies or supervises the application of a restricted-use pesticide on land owned or rented by him or his employer for the purpose of producing an agricultural commodity.

Private Commercial Applicator, annual license fee $23: A person who applies or supervises the use of a restricted-use pesticide on land owned or rented by him or his employer for purposes other than the production of an agricultural commodity.

Public Operator, annual license fee $23: A person who, while acting as an employee of a governmental agency, applies restricted-use pesticides by any means or general-use pesticides by power equipment on public or private property. Public operators may act as public consultant.

Public Pest Control Consultant, annual license fee $21: A person who, while acting as an employee of a governmental agency, offers or supplies technical advice, supervision, aid or makes recommendations to the user of pesticides other than home and garden products. Public Consultants may not act as public operators without the operator's license.

Demonstration and Research Applicator, annual license fee $23: A person who applies or supervises the use of a restricted-use pesticide to small experimental plots at no charge. Public employees performing research applications fall under the licensing requirements of the public operator.

License fees include a $6 surcharge to fund the state's Pesticide Incident Reporting and Tracking review panel, formed by the state legislature to deal with pesticide health-related issues and investigations.

To become licensed, individuals must receive a passing grade of 70 percent in all categories in which they will be working. In addition, passage of the laws and safety exam is necessary for all licenses except the private applicator and dealer manager.

Exams are given in a wide array of topics, approximately 25 in all. The category exams include the usual agricultural weed control and PCO structural. But there are also ones for stored grain, structural and turf demossing, methyl bromide, sprout inhibitor (potato storage), pole fume and birds on bridges.

After five years each license holder wanting to continue the license must apply for recertification. This can be accomplished through credit accumulation at WSDA-approved courses or retesting. The number of credits required is based on the license type. For private applicators, 20 hours of credit are required; all other license types must have 40 hours.

Washington requires licensing of consultants and of dealers who distribute non-home and garden pesticides.

It also requires that records be kept of applications by all certified applicators and all individuals applying pesticides to more than an acre of agricultural land per year.

Washington also has its own list of restricted-use pesticides for groundwater protection reasons. Most aquatic pesticides are state restricted-use as are certain formulations of 2,4-D in counties east of the Cascades.

It also requires that signs displaying the name and telephone number of the applicator be placed on apparatus being used on landscape and right-of-way applications.

Washington also has special restrictions on vertebrate control pesticides such as Compound 1080 and tributyltin.

The state has posting regulations for ornamental applications and a system for notifying pesticide-sensitive people.

WEST VIRGINIA

West Virginia requires private applicators to take an exam to be certified in either an agricultural pest control category or one for agricultural fumigation.

The state certifies commercial applicators with 13 categories. It also certifies public employees as certified public applicators.

There is also provision for registered technicians. They are non-certified employees of a pesticide business performing pesticide application similar to a commercial applicator or certified public applicator. Must complete a state-approved training program prior to issuance of an identification card, which costs $10.

West Virginia requires pesticide application businesses to pay a $50 annual fee.

Pesticide consultants must get a license and pay a fee of $50.

Pesticide dealers who sell restricted-use pesticides must get a license for $10.

To be recertified, West Virginia requires that at three year intervals each commercial applicator and certified public applicator take at least 20 continuing certification units, or 10 hours of training.

Private applicators must have 10 units to be recertified.

There are also requirements for recordkeeping by commercial applicators and certified public applicators.

WISCONSIN

Wisconsin gives private applicators an open book test in which they must get at least 50 percent of the answers right.

Commercial for-hire applicators who apply any pesticides must take self-study tests and get 70 percent or more right. Commercial applicators not-for-hire must take the test if they use restricted-use pesticides.

The cost to private applicators is $20 including the manual and study guide.

Commercial applicators are charged $30, plus $10 for the test and $50 for a two-year license.

Recertification happens every five years with an exam.

Wisconsin also regulates atrazine and metam sodium.

FLEA FACTS

-- More than 2,400 species exist worldwide.
-- A female flea can lay as many as 40 eggs per day and up to 400 in her lifetime.
-- A female flea consumes 15 times its body weight in blood daily.
-- Fleas can jump up to 150 times the length of their bodies -- sideways or up -- equivalent to a man jumping nearly a thousand feet.

From an Atlanta Journal - Constitution series on pesticide resistance

WYOMING

Wyoming does not require private applicators to take an exam to be certified if they attend a four hour training seminar or complete a home study workbook.

Commercial applicators, however, must take the exam.

The certification is good for three years for commercial applicators and five years for private.

The state charges $10 per category, with a maximum of $50, for the exam for commercial applicators. There is also an optional study manual for $25.

Only those commercial and private applicators who use restricted-use pesticides must be certified. That may be expanded to general-use pesticides in the future, the state's regulatory staff said.

Wyoming has 10 categories for certification of commercial applicators.

There is no insurance requirement for commercial applicators in Wyoming, although they are not released from any liability.

The agricultural commissioner may exempt any person from the licensing provisions of the law if he determines that emergency conditions exist, such as pest damage about to occur with no general-use pesticide available or if significant economic or health problems will occur without the use of the restricted-use pesticide.

Pesticide dealers must be licensed. There is no fee.

Dealers and certified applicators must keep records of restricted-use pesticides they sell and use for two years.

"When you try to eradicate an insect, you are going up against a billion years of evolution. Pests have survived that long because they are very good at adapting. We will probably never completely eradicate any pest. We shouldn't be trying. We should be looking for a way to live with them better."

Robert Metcalf, University of Illinois entomologist, quoted in the Atlanta Journal-Constitution

Chapter 8

CALIFORNIA

California has always been known as a leader in environmental regulation, and the pesticides area is no exception. The state has a larger pesticide bureaucracy than the U.S. Environmental Protection Agency's. It even has its own program to review the health effects of pesticides independent of the EPA, requiring manufacturers to submit most of the same studies to the state that they supply to EPA to support their federal registrations — and more speedily than EPA wants them.

The state also has its own unique rules on protection of air, surface and groundwater; reporting of pesticide use; worker health and safety; investigation of worker poisonings; and development of alternatives to pesticides.

There are reasons for that. California has the largest agricultural output of any state and uses the most pesticides. It also uses a large number of different pesticides because of the more than 250 different commodities grown in the state.

There's also the political climate in California, where voters have placed a high premium on environmental protection.

In the past few years California's pesticide regulatory program has taken on even more of a stringent regulatory stance with its inclusion into the newly formed California Environmental Protection Agency, known as Cal-EPA. Before, it was included in the Department of Food and Agriculture. Pesticide use is also regulated by the state's Structural Pest Control Board.

California's system is important even to those who don't work under its rules because it is the standard against which other states compare their pesticide programs. Eventually, many observers say, all states will have rules as stringent as those in California.

HOW THE SYSTEM AFFECTS USERS

Enforcement of the state rules for all types of pesticide use — from farmers to structural pest control workers to landscape contractors — is carried out by inspectors in the county agricultural commissioners offices in each county in the state.

By law, pest control businesses, agricultural pest control advisers and pest control aircraft pilots must register with each county in which they operate.

Anyone purchasing a restricted-use pesticide comes into contact with this system because they are required to get a permit from the agricultural commissioner's office. This has traditionally meant farmers, pest control companies and agricultural consultants. But the state is trying to pull in other people such as landscape gardeners who have not always been in the regulatory loop.

County inspectors inspect the operations and records of growers, pest control operators, pesticide dealers and agricultural pest control advisers;

register licensed pest control businesses, pest control aircraft pilots and agricultural pest control advisers; conduct pesticide incident investigations; provide training to pesticide users; and collect fresh produce samples for state pesticide residue monitoring programs.

As do other states, California certifies applicators of restricted-use pesticides. But it also licenses pest control operators, agricultural aircraft pilots, pesticide dealers and advisers. The licensing requirements date back as far as 1947, when herbicide drift problems in the San Joaquin Valley prompted the state legislature to require the licensing of agricultural pest control businesses.

When pesticides are used, applications must be reported to the state. While this system has been in place for decades, it's recently been expanded to include farmers and golf course workers, among others.

Under the system, all agricultural pesticide use must be reported in documents submitted monthly to the county agricultural commissioner. The reports must be site-specific and detail the kind and amount of pesticides used on specific commodities, according to a 1991 report describing the program to the state legislature.

Because California's definition of "agricultural use" is broad, it includes applications made to parks, golf courses, cemeteries and along roadsides and railroad right-of-ways so these applications must be reported. Also, all post-harvest pesticide treatments of agricultural commodities must be reported, along with all pesticide treatments in poultry and fish production, as well as some livestock applications. Home-use pesticides are exempt.

Also, under separate groundwater laws, the use of certain pesticides known to contaminate groundwater must be reported. This includes not only agricultural pesticides, but applications of certain pesticides in outdoor institutional or industrial situations.

Also, 24-hour notice of intent must be given to the county agriculture department when restricted-use pesticides are applied.

California's Department of Pesticide Regulation within Cal-EPA compiles the information from this reporting system in a number of ways to evaluate potential risks from pesticides in food, investigate potential misuses and farmworker illnesses; and develop exposure assessments for sensitive sites.

To ensure that all the rules are carried out, inspectors from the agricultural commissioners' offices conduct thousands of compliance inspections each year. They take place at use and mixing and loading sites, where pesticides are sold and stored and at locations where state-mandated records are kept.

A copy of the pesticide use inspection report is included here. It includes a long list of items for inspectors to go through, including the site, the application itself, equipment, field worker safety and mixing/loading.

Chris Browning, an inspector for San Luis Obispo County's agricultural commissioner's office, explained that the average inspection takes about an hour and a half.

He told a recent pesticide safety meeting that California inspectors have the authority to stop the application of pesticides at a site being inspected if the inspector believes an applicator's safety is in danger or the chemical is likely to drift.

He further explained some of the rules applicators must comply with in the state, some of which apply only to use of restricted materials:

—The pesticide label must always be at the site of application, even if that is different from where the product was mixed and loaded into equipment.

—Applicators must be trained to understand the label requirements on the pesticide being used, and must be sure that training is on the most current version of the label, which change occasionally.

—If the label specifies that a face shield be worn and only goggles are being used, that would be a violation.

—If a toxicity 1 category pesticide is being used, the worker must wear clean clothing each day and the employer must launder it. There must also be an extra set of clean clothes for the applicator at the site.

Also included at the end of this chapter are copies of some forms discussed here, such as the notice of intent to apply restricted materials and the monthly use report.

Pesticide applicators in California must also keep records of the training they have done of their workers. That form is also included.

In an attempt to improve compliance with worker safety regulations, the Department of Pesticide Regulation has also added a new set of "crop sheets" in English and Spanish explaining in simple terms the regulations that affect applicators. The crop sheets are meant to be kept in sight of the workers. A copy is included.

Table 3

STATE OF CALIFORNIA
PESTICIDE USE MONITORING INSPECTIONS
33-021 (REV. 8/90)

DEPARTMENT OF FOOD AND AGRICULTURE
PESTICIDE ENFORCEMENT BRANCH

PAGE _____ OF _____

FIRM/PERSON INSPECTED	BUSINESS TYPE/LICENSE NUMBER	INSPECTING COUNTY
FIRM ADDRESS	PHONE NUMBER	TIME STARTED/ENDED
APPLICATOR'S NAME	APPLICATOR'S LIC./CERT. NUMBER	EQUIPMENT ID NUMBER
PROPERTY OPERATOR	PERMIT/OPERATOR ID NUMBER	SITE ID NUMBER
PROPERTY LOCATION/ADDRESS	COMMODITY/SITE	PEST CONTROL ADVISER'S NAME
ADJACENT ENVIRONMENT (N) (S) (E) (W)		WIND VELOCITY/DIRECTION

PESTICIDE NAME/MANUFACTURER	LABEL REGISTRATION NUMBER	CAT.	FORM.	METHOD	RATE	DILUTION

A. PREAPPLICATION SITE INSPECTION	REF. SECTION	COMPLIANCE		
		YES	NO	N/A
1. Notice of Intent Consistent with Permit	6434			
Permit Monitoring				
2. - Proposed Applic. Complies w/ Permit Cond.	6432, 36			
3. - Environ. Cond. Consistent w/ Permit NOI	6436			
4. - Written Recommendation Reviewed	6436			
Total				

B. APPLICATION INSPECTION	REF. SECTION	COMPLIANCE		
		YES	NO	N/A
1. Registered Label Available at Use Site	6602			
Complies with Required Labeling	12973			
2. - Site/Rate/Concentration/Other				
3. - Protective Clothing				
4. - Chemical Resistant Gloves				
5. - Eye Protection				
6. - Chemical Resistant Boots				
7. - Respiratory Protection				
8. - Employee(s) Trained	6724			
9. - Emergency Medical Care Posting	6726			
10. Employee(s) Supervised, Category I	6730			
11. Clean Work Clothing, Categories I, II	6736			
12. Eye Protection Provided/Used	6738(a,b)			
13. Chemical Resistant Gloves Provided/Used	6738(a,c)			
14. Chemical Resistant Clothing Provided/Used	6738(a,d)			
15. Respiratory Equipment Provided/Used	6738(a,e)			
16. Complies with Permit Conditions	14007			
17. Notice of Intent Submitted	6434			
18. Restricted Materials Use Supervised	6404, 06			
19. Suitable Methods/Manner/Climate	6600			
20. Protection of Persons/Animals/Property	6614			
21. Drift Control/Phenoxy Herbicide Specifications	6460, 64			
22. Warning Signs Posted	6770, 76			
Total				

C. EQUIPMENT INSPECTION	REF. SECTION	COMPLIANCE		
		YES	NO	N/A
1. Equipment Registered in County	11732			
2. Equipment Identified	6630			
3. Equipment Safe to Operate	6600			
4. Uniform Mixture	6606			
5. Backflow Prevention Available	6610			
Safe Equipment	6742			
6. - Proper Tank Cover				
7. - Shut-Off Device/Sight Gauge				
8. Service Containers Labeled	6678			
Total				

D. FIELD WORKER SAFETY INSPECTION	REF. SECTION	COMPLIANCE		
		YES	NO	N/A
Field Supervisor Informed/Knowledgeable	6764			
1. - What Posting Means				
2. - Activities Prohibited During Reentry				
3. - Symptoms OP/Carbamate Poisoning				
4. - Emergency Medical Care Knowledge	6766(b)			
5. Hand Washing Facilities	6768			
Field Reentry After Pesticide Applied	6770(c)			
6. - Safe Reentry/Interval Expired				
7. Allowable Activities/6772(a) List Materials	6770(d)			
Conditional Reentry/6772(b) or by Label	6770(c)			
8. - Employee(s) Informed				
9. - Proper Protective Clothing				
10. - Employee(s) Instructed to Shower				
11. Posting Compliance	6776(a)			
Total				

E. MIX/LOAD INSPECTION	REF. SECTION	COMPLIANCE		
		YES	NO	N/A
1. Registered Label Available at Use Site	6602			
Complies with Required Labeling	12973			
2. - Site/Rate/Concentration/Other				
3. - Protective Clothing				
4. - Chemical Resistant Gloves				
5. - Eye Protection				
6. - Chemical Resistant Boots				
7. - Respiratory Protection				
8. - Employee(s) Trained	6724			
9. - Emergency Medical Care Posting	6726			
10. Employee(s) Supervised, Category I	6730			
11. Washing Facilities, Categories I, II	6734			
12. Work Clothing/Extra Clothing, Categories I, II	6736			
13. Eye Protection Provided/Used	6738(a,b)			
14. Chemical Resistant Gloves Provided/Used	6738(a,c)			
15. Chemical Resistant Clothing Provided/Used	6738(a,d)			
16. Respiratory Equipment Provided/Used	6738(a,e)			
17. Restricted Material Use Supervised	6404, 06			
18. Accurate Measurement	6604			
19. Closed System Used	6746			
20. Containers Secure/Under Control	6670			
21. Containers Properly Rinsed	6684			
Total				

REMARKS: (USE INSPECTION REPORT SUPPLEMENT IF ADDITIONAL SPACE IS NEEDED)

FOLLOW UP REQUIRED	NOTICE OF VIOLATION	CEASE AND DESIST	The noncompliance items noted above are violations and must be corrected
☐ YES ☐ NO	☐ YES ☐ NO	☐ YES ☐ NO	by:

ENFORCEMENT OFFICIAL SIGNATURE	DATE INSPECTED	INSPECTION ACKNOWLEDGEMENT	DATE ACKNOWLEDGED

Distribution: Original - County   First Copy - Inspector   Second Copy - Person/Firm Inspected

Table 3

STATE OF CALIFORNIA
DEPARTMENT OF FOOD AND AGRICULTURE
PRODUCTION AGRICULTURE MONTHLY PESTICIDE USE REPORT

Check this box only if you are a nursery.

MONTH: January YEAR: 1991 NURSERY: ☐

Page 1 of 1

OPERATOR ID/PERMIT NO.: 40-91-4019999
OPERATOR (GROWER): Joseph Grower
ADDRESS: 123 Crop Way
CITY: San Luis Obispo
ZIP CODE: 93401

SITE ID NO.: R1P001CA
TOTAL PLANTED ACREAGE/UNITS: 20
COUNTY NO.: 40
SECTION: 35
TOWNSHIP: 30
RANGE: 14 12 E W
BASE & MERIDIAN: 19 S M H

COMMODITY/SITE TREATED: Strawberries

FIELD LOCATION: (Optional)

CHEMICAL NUMBER 21	DATE/TIME APPLICATION COMPLETED 22	ACREAGE/UNITS TREATED 23	APPL. METHOD (CHECK ONE) 24	BLOCK ID IF APPLICABLE 25	EPA/CALIF. REG. NO. FROM LABEL 26	TOTAL PRODUCT USED (CIRCLE ONE UNIT OF MEASURE) 27	DATE REMEDY 28	RATE PER ACRE 29	DILUTION 30	PRODUCT/MANUFACTURER 31
	Jan. 8 7:30 AM	19	GR☐ AIR☐ OTHER☐		352-447	19 LB (OZ) PT QT GA	1	1 lb	100 gal.	Benlate 50 DF (DuPont)
	Jan. 20 7:30 AM	20	GR☐ AIR☐ OTHER☐		279-739-ZA	50 (OZ) PT QT GA LB	N/A	25 lb	100 gal.	Malathion 25 (FMC)
	Jan. 28 9:00 AM	15	GR☐ AIR☐ OTHER☐		3125-50001	75 LB (OZ) PT QT GA	1	5 lb	100 gal.	Dyrene 50% WP (Mobay)
			GR☐ AIR☐ OTHER☐			LB OZ PT QT GA				
			GR☐ AIR☐ OTHER☐			LB OZ PT QT GA				
			GR☐ AIR☐ OTHER☐			LB OZ PT QT GA				
			GR☐ AIR☐ OTHER☐			LB OZ PT QT GA				
			GR☐ AIR☐ OTHER☐			LB OZ PT QT GA				
			GR☐ AIR☐ OTHER☐			LB OZ PT QT GA				
			GR☐ AIR☐ OTHER☐			LB OZ PT QT GA				

Example

REPORT PREPARED BY: Joseph Grower   DATE: 2-8-91   REVIEWED BY: ____

Submit to Agricultural Commissioner within 10 days of month following application

33-017C (3/90)

(1) CAC   For Agency Use Only

Table 3

EMPLOYEE PESTICIDE SAFETY TRAINING RECORD

EMPLOYEE NAME:

EMPLOYER NAME:

TRAINER NAME:

EMPLOYEE'S SIGNATURE:

EMPLOYER'S SIGNATURE:

TRAINER'S SIGNATURE:

☐ APPLICATOR   ☐ FLAGGER

☐ MIXER/LOADER   ☐ OTHER (SPECIFY) _____

IMPORTANT:

1. Training must be given before employees are allowed to handle any pesticide, continually updated to cover any new pesticides that will be handled, and repeated and documented at least annually thereafter.

2. Training record must be retained for two years at a central location at the workplace accessible to the employee.

PESTICIDE NAME

1. IMMEDIATE AND LONG TERM HAZARDS involved, including hazards associated with exposure to pesticides known or suspected of chronic effects such as, tumors, cancer, birth defects, etc.

2. SAFETY PROCEDURES to be followed while mixing, loading, applying pesticides, or servicing contaminated equipment.

3. ENGINEERING CONTROLS: When and how to use enclosed cabs, closed mixing/loading systems

4. PROTECTIVE CLOTHING AND EQUIPMENT: Proper use and care of coveralls, gloves, goggles, boots, apron, rainsuit, respiratory equipment.

5. EMERGENCY PROCEDURES to be followed for handling non-routine tasks or emergency situations.

6. WAYS POISONING OR INJURY CAN OCCUR through ingestion, inhalation, or dermal routes.

7. IMPORTANCE OF IMMEDIATE DECONTAMINATION OF SKIN AND EYES when exposure occurs.

8. COMMON SYMPTOMS OF PESTICIDE POISONING: Pinpoint pupils, nausea, blurred vision, shortness of breath, dizziness, headache.

9. WHERE TO OBTAIN EMERGENCY MEDICAL TREATMENT: Name, address and telephone number of medical facility where emergency medical care is available.

10. PURPOSE AND REQUIREMENTS OF MEDICAL SUPERVISION: Required when handling pesticides with "Danger" or "Warning" signal word that contains an organophosphate or carbamate.

11. APPLICABLE LAWS AND REGULATIONS; MSDSs; PSISs; AND LABEL REQUIREMENTS: Importance of complying with the laws and regulations and label requirements.

12. EMPLOYEE RIGHTS: To personally receive information on pesticides they may be exposed to; to have physician or their representative receive this information; protected against discharge or discrimination.

13. LOCATION OF DOCUMENTS/RECORDS: Location of written Hazard Communication Program, use records, PSISs, MSDSs, exposure and monitoring records.

* ENTER DATE TRAINING GIVEN FOR EACH PESTICIDE (Month, day, year)

9/92

Table 3

CALIFORNIA CROP SHEET

Location of Records and Safety Information

Name
Telephone #

Address

Emergency Medical Care

Facility Doctor:
Telephone #

Address:

FLOWERS

SAFETY TIPS

- Protect your skin
- Wear clothes with long sleeves, long pants
- Wear shoes or boots, socks, a hat and/or scarf and gloves
- Make sure they are clean and without holes

- Take a bath or shower after work
- Wash with soap and shampoo
- Put on clean clothes

- Always wash your hands before eating, drinking, smoking or going to the bathroom
- Do not cook food with wood found in the field

- Pesticides get on work clothes and then on skin
- Wash work clothes before wearing again
- Wash work clothes separate from other clothes

- Never put pesticides in food containers
- Do not take farm pesticides home
- Keep children away from pesticides

WORKER RIGHTS

- You have the right to know the pesticides used where you work
- The information can be found at the address above (or in Pesticide Safety Information Series A9)
- You can call and ask for the spray record or go look at it

- You can confidentially file a complaint of unsafe conditions
- Call the County Agricultural Commissioner or Cal/OSHA
- The phone number is in the government pages of the phone book
- You cannot be fired for filing a complaint

ILLNESS/INJURY

- Tell the boss if you are sick or hurt
- The first doctor's visit will be paid by your employer
- If the illness/injury is work related, your employer will pay for all medical care of that illness and time off work while sick or hurt

REENTRY INTERVALS

DANGER   PELIGRO

PESTICIDES NO ENTRE
DO NOT
ENTER

- Pesticides get on your skin and clothes when you touch sprayed plants
- Pesticides take time to go away
- Reentry intervals keep you out of the field until it is safe

- Signs like this tell you to stay out of the field
- Irrigators & tractor drivers may go into posted fields if they are told about the dangers and safe work procedures and are using the protection required

FIRST AID

CALIFORNIA

Regions

- Wash immediately if sprayed or exposed to spray drift
- Change into clean clothes
- Tell your boss after washing

- Wash if your eyes or skin begin to itch or burn
- Use lots of water
- Tell your boss, you should go to a doctor

- If you feel sick (headache, stomach ache, vomiting, dizzy, etc.) tell your boss
- He can make sure you are taken to a doctor

- Never drive yourself to the doctor if you are sick or injured

*CALIFORNIA CROP SHEET
PAGE 2
SEPTEMBER 1992*

FLOWERS

The pesticides listed are those most commonly used on this crop. These pesticides may or may not have been used on the field you are working in. You can ask your boss or contact the person on the front of this sheet (or in Pesticide Safety Information Series A9) for specific information.

Pesticides Used Most Often	J	F	M	A	M	J	J	A	S	O	N	D	Reentry Interval for Flowers	Common Symptoms of Overexposure	Notes
Methyl bromide Region 1 Region 2 Region 3													Application made to soil before planting	Headache, nausea, dizziness, vomiting and tremors. Severe burning, itching and blisters on the skin	If you have any questions, you can contact your Agricultural Commissioner's office or the Worker Health and Safety Branch (Cal/EPA, Department of Pesticide Regulation), 1220 N Street, Suite 620, Sacramento, California 95814 (916) 654-0445
Chloropicrin Region 1 Region 2 Region 3													Application made to soil before planting	Severe irritation of the respiratory system, skin and eyes	
Metam sodium Region 1 Region 2 Region 3													Application made to soil before planting	Burning or itching skin or eyes. Respiratory irritation (coughing, sneezing, hard to breathe)	
Sulfur Region 1 Region 2 Region 3													Spray is dry Dust is settled	Burning or itching eyes or skin. Respiratory irritation (coughing, sneezing, hard to breathe)	
Piperalin Region 1 Region 2 Region 3													Spray is dried Dust is settled	Skin and eye irritation have been reported in California	
Copper sulfate Region 1 Region 2 Region 3								No use No use				Spray is dried Dust is settled	Irritating to the eyes, skin and respiratory tract	

Developed by: California Environmental Protection Agency, Department of Pesticide Regulation, Worker Health and Safety Branch

Table 3

GUIA DE CULTIVOS DE CALIFORNIA

Localización de los Datos e Información de Seguridad

| Nombre | Teléfono # |
| Dirección | |

Servicio Médico de Emergencia

| Clínica o Médico: | Teléfono # |
| Dirección: | |

FLORES

CONSEJOS DE SEGURIDAD

- Proteja su piel
- Use manga larga y pantalones largos
- Use zapatos o botas, calcetines, un sombrero o una bandana y guantes
- Asegúrese que estén limpios y sin roturas

- Dése un baño o una ducha después del trabajo
- Lávese con jabón y champú
- Póngase ropa limpia

- Lávese siempre las manos antes de comer, beber o fumar, y antes de ir al baño
- No cocine con leña que ha encontrado en el campo

- Los pesticidas se adhieren a la ropa de trabajo y después pasan a la piel
- Lave la ropa de trabajo antes de volverla a usar
- Lave la ropa de trabajo separada de otra ropa

- Nunca ponga pesticidas en envases de alimentos
- No lleve a casa los pesticidas que usa en el campo
- Mantenga los niños alejados de los de pesticidas

DERECHOS DE LOS TRABAJADORES

- Usted tiene derecho a saber que pesticidas se usan en su trabajo
- Esa información puede encontrarla en la dirección ya mencionada (o en la serie informativa Seguridad con Pesticidas A-9)
- Usted puede pedirla por teléfono o puede ir a mirarla

- Denuncias de condiciones peligrosas son confidenciales
- Llame al Comisionado Agrícola del Condado o Cal/OSHA
- Número telefónico en sección gobierno de la guía
- No pueden hecharlo del trabajo por presentar una denuncia

ENFERMEDAD/ LESION

- Avísele al jefe si está enfermo o se ha lesionado
- La primera visita al médico la pagará el empleador
- Si la enfermedad o lesión es a causa del trabajo su empleador pagará toda la atención médica y el tiempo que no pueda trabajar

INTERVALO DE ENTRADA

DANGER	PELIGRO
PESTICIDA	
DO NOT ENTER	NO ENTRE

- Los pesticidas se adhieren a su piel y ropa cuando toca plantas rociadas
- Los pesticidas toman tiempo en disiparse
- Los intervalos de entrada no le permiten entrar al campo hasta que pueda hacerlo sin peligro

- Letreros como este advierten no entrar al campo
- Los regadores o tractoristas pueden entrar siempre que se les advierta el peligro y los procedimientos de seguridad y que usen la protección requerida

PRIMEROS AUXILIOS

- Lávese inmediatamente si ha sido salpicado y expuesto al rocío
- Cámbiese a ropa limpia
- Después de lavarse avísele a su jefe

- Lávese si siente picazón o quemazón en los ojos o la piel
- Use bastante agua
- Avísele a su jefe que debería ir al médico

- Si siente malestar (dolor de cabeza, dolor de estómago, vómito, mareo, etc.) avísele a su jefe
- El se encargará de que lo lleven a un médico

- Nunca maneje usted si está enfermo o lesionado

GUIA DE CULTIVOS DE CALIFORNIA PAGINA 2 SETIEMBRE 1992

FLORES

Los pesticidas en la lista que sigue son los que se usan con más frecuencia en este cultivo. Estos pesticidas pueden haberse usado o no haberse usado en el campo donde usted está trabajando. Puede preguntarle a su jefe o ponerse en contacto con la persona mencionada al principio de este documento (o en la serie informativa Seguridad con Pesticidas A-9) para obtener información específica.

Pesticidas Usados con Mayor Frecuencia	Meses en que se Usan Generalmente E F M A M J J A S O N D	Intervalo de Entrada para Flores	Síntomas Comunes de Sobre-exposición	Notas
Metilo, Bromuro de Región 1 Región 2 Región 3		Aplicación al suelo antes de plantar	Dolor de cabeza, náusea, vértigos, vómito y temblor. Quemazón cutánea severa, picazón y ampollas en la piel	Si tiene alguna pregunta, puede comunicarse con la oficina del Comisionado de Agricultura de su localidad o la Sección de Salud del Trabajador (Cal/EPA, Departament de Reglamentación de Pesticidas), 1220 N Street Sacramento, California 95814 (916) 654-0445
Cloropicrin Región 1 Región 2 Región 3		Aplicación al suelo antes de plantar	Irritación severa al sistema respiratorio, cutánea y ojos	
Metam, Sodio de Región 1 Región 2 Región 3		Aplicación al suelo antes de plantar	Quemazón o picazón cutánea o los ojos Irritación respiratoria (tos, estornudos, dificultad en respirar)	
Azufre Región 1 Región 2 Región 3		El rocío se ha secado El polvo se ha asentado	Quemazón y picazón cutánea o los ojos Irritación respiratoria (tos, estornudos, dificultad en respirar)	
Piperalin Región 1 Región 2 Región 3		El rocío se ha secado El polvo se ha asentado	En California se han encontrado casos de irritación a la piel y en los ojos	
Cobre, Sulfato de Región 1 Región 2 Región 3		El rocío se ha secado El polvo se ha asentado	Irritación de los ojos, cutánea y vía respiratoria	

Preparado por: Agencia de Protección Ambiental de California Departamento de Reglamentación de Pesticidas Sección de Salud y Seguridad del Trabajador

DIRECTORY OF STATE OFFICIALS

Alabama
J.A. Bloch, director
Plant Protection and Pest Management Division
Alabama Department of Agriculture and Industries
P.O. Box 3336, 1445 Federal Drive
Montgomery, AL 36109-0336
Phone: (205) 242-2656

Alaska
Kit Ballentine, acting director
Environmental Health
Department of Environmental Conservation
Pouch O
Juneau, AK 99811
Phone: (907) 465-2696

Arizona
Janet E. Bessey, assistant director
Registration and Licensing
Arizona Department of Agriculture
1688 West Adams, Rm. 436
Phoenix, AZ 85007
Phone: (602) 542-0949

structural:
Jack Root, executive director
Structural Pest Control Commission
1150 S. Priest Drive, No. 4
Tempe, AZ 85281
Phone: (602) 255-3664

Arkansas
Charles Armstrong, assistant director
Division of Feeds, Fertilizer and Pesticides
Arkansas State Plant Board
P.O. Box 1069
Little Rock, AR 72203
Phone: (501) 225-1598

structural:
Kiven Stewart, head
Pest Control Section
Arkansas State Plant Board
P.O. Box 1069
Little Rock, AR 72203
Phone: (501) 225-1598

California
Mac Takeda, program specialist
Pesticide Use Enforcement and Licensing
California Department of Pesticide Regulation
1220 N Street, Room A-170
P.O. Box 942871
Sacramento, CA 94271-0001
Phone: (916) 654-0606

structural:
Mary Lynn Ferreira
registrar and executive secretary
Structural Pest Control Board
California Department of Consumer Affairs
1422 Howe Avenue, Suite 3
Sacramento, CA 95825
Phone: (916) 920-6089

Colorado
Linda Coulter, chief
Pesticide Section
Division of Plant Industry
Colorado Department of Agriculture
700 Kipling St., Ste. 4000
Lakewood, CO 80215-5894
Phone: (303) 239-4140

Connecticut
Linda Schmidt
environmental analyst
Pesticide Management Division
Department of Environmental Protection
165 Capitol Avenue
Hartford, CT 06106
Phone: (203) 566-5148

Delaware
W. Larry Towle
agricultural specialist
Delaware Dept. of Agriculture
2320 South DuPont Highway
Dover, DE 19901
Phone: (302) 739-4811

District of Columbia
Mark Greenleaf, branch chief
Pesticide Enforcement and Certification Branch
Environmental Regulation Administration
Department of Consumer and Regulatory Affairs
2100 Martin Luther King Jr. Ave SE, Ste. 203
Washington, DC 20020
Phone: (202) 404-1167

Florida
Elisabeth Braxton, administrator
Compliance Section, Bureau of Pesticides
Division of Agricultural Environmental Services
Florida Department of Agriculture and Consumer Services
3125 Conner Boulevard, MC-1
Tallahassee, FL 32399-1650
Phone: (904) 488-3314

structural:
Dr. John A. Mulrennan Jr.
Bureau of Entomology and Pest Control
Florida Dept. of Agriculture and Consumer Services
P.O. Box 210
Jacksonville, FL 32231
Phone: (904) 798-4594

Georgia
Tommy Gray, ag manager I
Entomology & Pesticide Division
Georgia Department of Agriculture
Capitol Square
Atlanta, GA 30334
Phone: (404) 656-4958

structural:
Jim Harron, ag manager II
Entomology & Pesticide Division
Georgia Department of Agriculture
Capitol Square
Atlanta, GA 30334
Phone: (404) 656-3641

Hawaii
Robert Boesch, program manager
Pesticides Branch
Plant Industry Division
Hawaii Department of Agriculture
1428 South King St.
Honolulu, HI 96814
Phone: (808) 973-9402

Idaho
Taylor Cox, certification specialist
Bureau of Agricultural Standards
Division of Agricultural Technology
Idaho Department of Agriculture
P.O. Box 790
Boise, ID 87301-0790
Phone: (208) 334-3243

Illinois
Sherri Powell, office administrator
Illinois Department of Agriculture
P.O. Box 19281
Springfield, IL 61794-9281
Phone: (217) 785-2427

structural:
Fred Riecks, manager
Pesticides Program
Division of Environmental Health
Illinois Department of Public Health
535 West Jefferson
Springfield, IL 62761
Phone: (217) 782-5830

Indiana
Carl R. Rew, manager
Certification and Licensing
Indiana State Chemist Office
1154 Biochemistry Building
Purdue University
West Lafayette, IN 47907-1154
Phone: (317) 494-1594

Iowa
Charles A. Eckermann, chief
Pesticide Bureau
Iowa Department of Agriculture
Wallace Building
Des Moines, IA 50319
Phone: (515) 281-8591

Kansas
Plant Health Division
Kansas Department of Agriculture
901 S. Kansas St.
Topeka, KS 66612-1281
Phone: (913) 296-2263

Kentucky
Ron Egnew, director
Division of Pesticides
Kentucky Department of Agriculture
500 Mero St., 7th floor
Frankfort, KY 40601
Phone: (502) 564-7274

Louisiana
Larry LeJeune, assistant director
Pesticide Use
Pesticide and Environmental Programs
Louisiana Department of Agriculture and Forestry
P.O. Box 3596
Baton Rouge, LA 70821-3596
Phone: (504) 925-3768

Maine
Henry S. Jenning, chief
Certification and Enforcement
Board of Pesticides Control
Maine Department of Agriculture
State House Station No. 28
Augusta, ME 04333
Phone: (207) 287-2731

Maryland
Mary Ellen Setting
Pesticide Regulation Section
MD Department of Agriculture
50 Harry S. Truman Parkway
Annapolis, MD 21401
Phone: 301-841-5710

Massachusetts
Mark S. Buffone, certification coordinator
Pesticide Branch
Massachusetts Dept. of Food and Agriculture
100 Cambridge St., 21st floor
Boston, MA 02202
Phone: (617) 727-3020

Michigan
Katherine Fedder
pesticide enforcement and certification manager
Pesticide and Plant Pest Management Division
Michigan Department of Agriculture
P.O. Box 30017
Lansing, MI 48909
Phone: (517) 335-6838

Minnesota
Wayne Dally
pesticide control specialist
Agronomy Services Division
Minnesota Dept. of Agriculture
90 West Plato Blvd.
St. Paul, MN 55107
Phone: (612) 297-2746

Mississippi
Tommy McDaniel
pesticide coordinator
Bureau of Plant Industry
Mississippi Dept. of Agriculture and Commerce
P.O. Box 5207
Mississippi State, MS 39762
Phone: (601) 325-3390

structural:
James Haskins
applicator certification
Bureau of Plant Industry
Mississippi Dept. of Agriculture and Commerce
P.O. Box 5207
Mississippi State, MS 39762
Phone: (601) 325-3390

Missouri
Paul Bailey, program manager certification
Bureau of Pesticide Control
Plant Industries Division
Missouri Department of Agriculture
P.O. Box 630
Jefferson City, MO 65102-0630
Phone: (314) 751-2462

Montana
George A. Algard, chief
Technical Services Bureau
Agricultural and Biological Sciences Division
Montana Department of Agriculture
Capitol Station
Helena, MT 59620-0205
Phone: (406) 444-2944

Nebraska
EPA Region VII handles private and commercial certification for Nebraska.
Jim Mulligan
Regional Toxics and Pesticide Branch
726 Minnesota Avenue
Kansas City, KS 66101
Phone: (913) 236-2800

Nevada
Charles Moses, pesticide specialist
Nevada Department of Agriculture
P.O. Box 11100, 350 Capitol Hill Ave.
Reno, NV 89510-1100
Phone: (702) 688-1180

New Hampshire
Murray L. McKay, director
Division of Pesticide Control
NH Dept. of Agriculture
Caller Box 2042
Concord, NH 03302-2042
Phone: (603) 271-3550

New Jersey
Ralph C. Smith, chief
Bureau of Pesticide Operations
Pesticide Control Program
NJ Dept. of Environmental Protection
CN 411 380 Scotch Road
Trenton, NJ 08625
Phone: (609) 530-4134

New Mexico
Lonnie Mathews, chief
Bureau of Pesticide Management
Division of Agricultural and Environmental Services
NM Department of Agriculture
P.O. Box 30005, Dept. 3AQ
Las Cruces, NM 88003
Phone: (505) 646-2133

New York
Marilyn M. DuBois, director
Bureau of Pesticide Regulation
NY Dept. of Environmental Conservation
50 Wolf Road
Albany, NY 12233-7254
Phone: (518) 457-7482

North Carolina
James W. Burnette, Jr.
assistant pesticide administrator
Food and Drug Protection Division
NC Dept. of Agriculture
P.O. Box 27647
Raleigh, NC 27611-0647
Phone: (919) 733-3556

structural:
N. Ray Howell, director
Structural Pest Control Division
NC Dept. of Agriculture
P.O. Box 27647
Raleigh, NC 27611-0647
Phone: (919) 733-6100

North Dakota
Jack Peterson, director
Pesticide Division
ND Dept. of Agriculture
600 E Blvd., 6th floor
Bismarck, ND 58505-0020
Phone: (701) 224-2231

Ohio
Robert Wulfhorst
specialist in charge
Pesticide Regulation
Division of Plant Industry
Ohio Department of Agriculture
8995 East Main St.
Reynoldsburg, OH 43068-3399
Phone: (614) 866-6361

Oklahoma
Sandy Wells, program manager
Pesticide Application Certification and Licensing Program
Plant Industry and Consumer Services Division
Oklahoma Dept. of Agriculture
2800 North Lincoln Blvd.
Oklahoma City, OK 73105-4298
Phone: (405) 521-3864 Ext. 273

Oregon
James L. Sandeno
registration/certification supervisor
Plant Division
Oregon Dept. of Agriculture
635 Capitol St. NE
Salem, OR 97310
Phone: (503) 378-3776

Pennsylvania
John Tacelosky
certification and education specialist
Division of Agronomic Services
Penn. Dept. of Agriculture
Bureau of Plant Industry
2301 North Cameron St.
Harrisburg, PA 17110-9408
Phone: (717) 787-4843

Rhode Island
Elizabeth M. Lopes-Duguay
senior plant pathologist
RI Dept. of Environmental Management
22 Hayes St.
Providence, RI 02908
Phone: (401) 277-2781

South Carolina
Neil Ogg, assistant department head
Department of Fertilizer and Pesticide Control
256 Poole Agricultural Center
Clemson, SC 29634-0394
Phone: (803) 656-3171

South Dakota
Brian Scott, supervisor
Pesticide Activity
Division of Regulatory Services
SD Dept. of Agriculture
Anderson Bldg., 445 East Capitol
Pierre, SD 57501-3185
Phone: (605) 773-3724

Tennessee
Knox Wright
Environmental Control Section
Division of Plant Industries
Tenn. Dept. of Agriculture
P.O. Box 40627, Melrose Station
Nashville, TN 37204
Phone: (615) 360-0130

structural:
Jace Burch
Structural Pest Control Section
Division of Plant Industries
Tenn. Dept. of Agriculture
P.O. Box 40627, Melrose Station
Nashville, TN 37204
Phone: (615) 360-0130

Texas
Lemarcus Johnson, special assistant
Pesticide Program
Texas Dept. of Agriculture
P.O. Box 12847
Austin, TX 78711
Phone: (512) 463-7526

structural:
Benny Mathis, executive director
Structural Pest Control Board
9101 Burnet Road, Ste. 201
Austin, TX 78758
Phone: (512) 835-4066

Utah
Gary King, supervisor
Pesticide and Fertilizer Inspection
Utah State Agricultural Dept.
350 North Redwood Road
Salt Lake City, UT 84116
Phone: (801) 538-7188

Vermont
Philip R. Benedict, director
Plant Industry
Laboratory and Standards Division
Vermont Dept. of Agriculture
120 State Street
State Office Building
Montpelier, VT 05620-2901
Phone: (802) 828-2431

Virginia
Marshall W. Trammell, Jr., supervisor
Education, Training and Information
Office of Pesticide Management
Virginia Dept. of Agriculture and Consumer Service
P.O. Box 1163, Rm. 403
Richmond, VA 23209
Phone: (804) 371-0152

Washington
Glenn E. Smerdon, program manager
Registration and Licensing Services
Washington Dept. of Agriculture
P.O. Box 42589
Olympia, WA 98504
Phone: (206) 902-2031

West Virginia
Robert Frame, director
Pesticide Division
West Virginia Dept. of Agriculture
Charleston, WV 25305
Phone: (304) 348-2209

Wisconsin
Ed Bergman, Certification and Licensing
Wisconsin Dept. of Agriculture, Trade and Consumer Protection
801 W. Badger Road, P.O. Box 8911
Madison, WI 53708
Phone: (608) 266-0197

Wyoming
Jim Bigelow, director
Technical Services
Wyoming Dept. of Agriculture
2219 Carey Ave.
Cheyenne, WY 82002-0100
Phone: (307) 777-6590

154